PRAISE FOR *INGREDIENTS*

"In a slyly brilliant bait and switch, what is framed as a book about what we should eat becomes a thriller about the scientific method itself. . . . Mr. Zaidan argues persuasively [that] the disagreements between nutritional epidemiologists and other scientists on this subject show science working as it should: It's messy and imperfect, but the struggle is a struggle toward truth."   —*The Wall Street Journal*

"I'm going to go out on a limb here and say that food is very important, and yet we are terrible at talking about it. Nutrition is a mess of marketing, classism, science, truth, guilt, confusion, and outright hucksterism. *Ingredients* lifts the film from our eyes with humor and reassurance."

—Hank Green, author of *An Absolutely Remarkable Thing*

"At last, a book on nutrition that tries to make you understand how little we know instead of offering blanket prognostications. If instead of a simple solution, you want a guide to how to think about health, this is it."

—Zach and Kelly Weinersmith, *New York Times*–
bestselling authors of *Soonish*

"If you are looking for a guide in understanding the everyday chemistry of our lives, you could not do better than George Zaidan. And his book, *Ingredients*, is everything that should lead you to expect: funny, edgy, fascinating, dismaying, reassuring, and, overall, just incredibly smart."

—Deborah Blum, Pulitzer Prize–winning author of
The Poison Squad

"By all means, pick up George Zaidan's high-octane *Ingredients* if you want to know more about Cheetos, sunscreen, butter substitutes, and other fascinating bits of everyday chemistry. But above all, you should buy *Ingredients* because it teaches you how to think better—like a smart, informed, and wickedly funny scientist."

—Sam Kean, author of *The Disappearing Spoon* and
The Bastard Brigade

"If you ever thought that chemistry might be really interesting (it is), but your eyes glazed over in high school chem class, this is the book for you. George Zaidan will keep you laughing out loud as he shares the wonders of our most useful, practical science, with brilliant analogies that even an eleven-year-old can understand."

—Daniel J. Levitin, *New York Times*–bestselling author of
Successful Aging and *This Is Your Brain on Music*

"Omfg this book is FABULOUS! It's hilarious, insightful, sassy, and reassuring. A delightful roller coaster of science communication."

—Kallie Moore, cohost of PBS's *Eons*

"George Zaidan's mix of razor-sharp wit and pinpoint accuracy is rarer in science than a T. rex performing nuclear fusion. *Ingredients* has the answers to age-old questions—How many Oreos are too many Oreos?—and many more you never thought to ask. Like an optometrist performing stand-up, Zaidan is eye-opening and hilarious."   —Daniel Stone, author of *The Food Explorer*

"Everything in our lives is made of chemicals. But unfortunately very few of us are chemists. *Ingredients* is a road map for navigating the confusing polysyllabic world we find in product labels and in viral news stories. Zaidan's blend of humor and science will not only make you a better-informed consumer of all things chemical. [It] will also make you appreciate the chemistry that makes our world possible."   —Joe Hanson, creator/writer/host of *It's Okay to Be Smart*

"Through incredibly weird and wonderful analogies (and delightfully nerdy wit), George helps you understand how scientists work toward the truth. I wish he'd rewrite all of my high school science textbooks!"   —Emily Calandrelli, author of the Ada Lace Adventures

"*Ingredients* is a friendly introduction to the chemistry behind our health, but it's also a compelling portrait of how science is conducted and knowledge is built. Turns out, Cheetos and the scientific method have something in common: there's a lot going on, and not everyone knows what. George does a masterful job of showing where chemistry can answer questions about our health and environment, and where it—as well as science in general—is led by politics, culture, and even *gasp* opinion."   —Mike Rugnetta, host of *Idea Channel*

"When I taught a writing-intensive course for nutrition and food science seniors, the main objectives were how to read scientific papers critically and how to argue effectively in print. I thought several times while reading this book that, rather than using peer-reviewed papers, I wish I could have had this book for my students. Pick any argument George makes and tell me, with references, why you agree or disagree. They probably would have learned more that way and certainly would have enjoyed their reading more."

—David Klurfeld, former professor and chairman of
the Department of Nutrition and Food Science
at Wayne State University

"*Ingredients* has all the ingredients I'm looking for in a science book: it's chock-full of interesting information; it reveals the science behind an everyday subject; it's written in a breezy, easy-to-understand voice—and it's funny! I can't recommend it enough."

—Brian Malow, science comedian

"Restrained, thoughtful, and eye-opening analysis . . . There is good information to be found in this book." —*Kirkus Reviews*

"An entertaining romp through the world of scientific studies focusing on topics that will concern most readers. Recommended for all curious about the everyday products they consume or use."

—*Library Journal*

"*Ingredients* employs a lighthearted tone and approachable language to enlighten even the least science-inclined reader on the strengths and pitfalls of the science that tells us what's best for our bodies." —*Booklist*

"If 'worry' is your default mode these days, get *Ingredients* and chill. You'll eat it up." —*Marco Eagle*

"From the start, George Zaidan's *Ingredients* distinguishes itself from the typical 'Eat this, not that' manifesto. . . . [It] is about the complex process of figuring out how to answer this question. In other words, instead of offering up faddish nutritional spin, or dishing out sanctimonious vagaries like, 'Eat real food,' Zaidan does something different—and much, much more worthwhile. . . . Zaidan has a gift for punching up hard science with goofball details without sacrificing substance." —*BookPage*

"In *Ingredients*, George Zaidan delivers an enthusiastic introduction to nutritional epidemiology. . . . Using simple illustrations and his trademark humor to demystify scientific analysis that doesn't always prove cause and effect, Zaidan empowers readers to make their own dietary decisions." —*Shelf Awareness* (starred review)

INGREDIENTS

THE STRANGE CHEMISTRY OF
WHAT WE PUT
IN US AND ON US

GEORGE ZAIDAN

ILLUSTRATED (POORLY)
BY THE AUTHOR

DUTTON

DUTTON

An imprint of Penguin Random House LLC

penguinrandomhouse.com

Previously published in a Dutton hardcover edition in April 2020

First Dutton trade paperback printing: April 2021

**DUTTON and the D colophon are registered trademarks
of Penguin Random House LLC.**

Library of Congress Cataloging-in-Publication Data
has been applied for.

Dutton trade paperback ISBN: 9781524744298

Printed in the United States of America
1   3   5   7   9   10   8   6   4   2

BOOK DESIGN BY LORIE PAGNOZZI

While the author has made every effort to provide accurate telephone numbers,
internet addresses, and other contact information at the time of publication, neither
the publisher nor the author assumes any responsibility for errors or for changes that
occur after publication. Further, the publisher does not have any control over and
does not assume any responsibility for author or third-party websites or their content.

Neither the publisher nor the author is engaged in rendering professional advice or
services to the individual reader. The ideas, procedures, and suggestions contained
in this book are not intended as a substitute for consulting with your physician. All
matters regarding your health require medical supervision. Neither the author nor
the publisher shall be liable or responsible for any loss or damage allegedly arising
from any information or suggestion in this book.

To Mom, Dad, and Julia:

Sorry.

CONTENTS

INGREDIENTS

PREFACE

Going to MIT was like going to Hogwarts. The place was full of witches and wizards doing stuff that was indistinguishable from magic. But the most magical part was suddenly finding myself among a bunch of fellow nerds—this was pre-Facebook, when nerds were still considered cute, harmless pets—and realizing that I was one of them. I could do magic, too.

I wish I had the bravery and recklessness of a Gryffindor, but I was a Ravenclaw through and through: quiet, weird, and never in trouble. In fact, my friends said I was "allergic to fun." In fairness, this was absolutely true. I spent most Friday nights working in my room, I can't recall going to a single party, and I voluntarily chose chemistry as a major, which meant that I took three semesters of organic chemistry (affectionately known as "orgo"). I then became a teaching assistant for that class . . . twice. So, yeah, definitely a severe fun allergy.

The best thing about intro organic chemistry is learning to build molecules—not in a lab, but on paper. You're given a few molecules to start with and the target molecule you have to build. Like this:

STARTING MATERIALS: BENZENE, FORMALDEHYDE
TARGET: DIPHENYLMETHANOL

Your job is to chart a path from the starting materials to the target. One answer to the above, for example, might be a five-step process involving iron bromide, bromine, magnesium, tetrahydrofuran, and pyridinium chlorochromate.

Okay, I realize that seems . . . like the opposite of magical. But learning this stuff is like taking a cooking class that teaches you how to invent new dishes, or forge knives, or create new cooking techniques—not just how to hold a knife or follow a recipe. Introductory orgo was just structured enough to make sense but just freeform enough to allow for creativity.

Then I took *advanced* orgo.

One day the professor came into class holding a Diet Coke. He took a long sip, tipping his head all the way back, exhaled an "Ahhhhhh" just like in the commercials, and then, as if mugging for a camera, proclaimed: "Diet Coke, the Elixir of Life." That wasn't unusual; he probably started half his lectures this way. (Strange man; great teacher.) As I remember it, he then wrote a chemical reaction on the board and asked us to predict the products:

$$\text{SOME CHEMICAL} + \text{SOME OTHER CHEMICAL} \rightarrow ?$$

I'd never seen the reaction before, and by the looks of everyone around me, they hadn't either. When no one answered, he added four letters:

$$\text{SOME CHEMICAL} + \text{SOME OTHER CHEMICAL} \rightarrow \text{AHBL}$$

"Anyone know what AHBL is?" he asked.

Thirty-seven lifelong overachievers immediately panicked. This hadn't been covered in previous semesters. And I hadn't rememorized the periodic table in years, but I was pretty sure AHBL was not on it. A and L are not elements; hydrogen (H) is not usually sandwiched between other atoms; and boron (B) usually takes three partners, not two. Plus it was weird that this was in all caps—

Oh.

| SOME CHEMICAL | + | SOME OTHER CHEMICAL | → | ALL HELL BREAKS LOOSE |

In other words, two fairly simple chemicals reacted to produce thousands of new products—completely and utterly useless to a chemist trying to cleanly synthesize one pure thing.

I think about that reaction to this day. On the left, simplicity. On the right, chaos. Overall, the exact opposite of the clean, magical reactions we all learned in intro organic chemistry.

There are so, so, *so* many different chemicals we put into our bodies every single day. Water. Cheetos. Cigarettes. Sunscreen. Vape mist. The list is almost literally endless. What happens when all that stuff interacts with all the chemicals that make up our bodies?

Does—in the immortal words of Professor Elixir of Life—AHBL?

If so, does all hell breaking loose impact our health?

I went searching for the answers, and I was surprised at what I found. Things out in Science Land were quite different from what I thought they'd be. But before we get to all that, I want to spend a bit of time on *how* I found the information I'm about to share.

I found it by reading.

Well, no shit, Sherlock.

I found it by reading *science*, which isn't reading so much as decoding or translating . . . because science really is a foreign language. It has its own special words, grammar, rhythms, slang, and even smackdowns. (For example, in English, describing someone as "not serious" just means they're fun or lighthearted; but in science those same words are a grievous insult, akin to whipping out your white glove and slapping someone in the face.)

Decoding science involves reading short publications that are intended only for other scientists. These are formally called "journal articles," but most scientists call them "papers." A paper is what scientists publish when they do an experiment they like—or have a thought they like—and want all the other scientists to know how awesome it was. This happens all the time, so there are a *ton* of papers out there: at least 60 million, with about two million new ones published *every year.* Learning to read these papers gives you access to—in Jasmine's words—a whole new world. If you have a question about how the world works—like "How do plants make sugar out of light and air?" or "What are the weirdest things people put in their butts?"—the first place you should look is the collection of all the papers in the world. Scientists call this "the literature."

So, to answer all the questions I had when writing this book, I turned to the literature. I read a few papers; I interviewed a few scientists. Then I read a few more papers and talked to other scientists. And then, as often happens when you plug into the literature, I got sucked in. When my tally reached a hundred papers, I realized that some facts I had previously learned were wrong. When it reached five hundred, I had found so many fascinating facts and interesting stories that I figured I should write about it. When my tally reached a thousand papers (and eighty interviews), I realized that I was looking at the world in a whole new way. I hope that you have the same experience reading this book as I did reading the literature.

Before we start our odyssey, though, let me be clear about who I am and what sights you can expect to see along the way. I am not a practicing scientist. For the past decade, my job has been to translate science into English as accurately and entertainingly as possible. So I don't mainline the literature like professional scientists do. I sip it, spit it out, and try to make sense of what I'm tasting—like a wine critic but with slightly less pomp and circumstance. So this book will inevitably contain mistakes. If you think you've found one, please let me know. You can e-mail me at oops@ingredientsthebook.com or hit me up on Twitter @georgezaidan and I'll dig into the mistake and see what I can discover.

And there's another caveat: because there is so much information out there, I had to leave lots of stuff on the cutting room floor. I've provided this handy chart so you know exactly what to expect—and not expect—from this book:

STUFF IN THIS BOOK	STUFF IN OTHER BOOKS
How bad is processed food? How sure are we?	Your carbon footprint
Is sunscreen safe? Should you use it?	Food sustainability
What about vaping?	GMOs
Is coffee good or bad for you?	Science funding
What's your disease horoscope?	Politics
What's that public pool smell made of?	Football
What happens when you overdose on fentanyl in the sun?	Baseball
What do cassava plants and Soviet spies have in common?	Any kind of ball, really
When will you die?	

All the topics on the right are important topics, and many of them are interwoven with topics on the left, but I need to save some material for books down the line.

Okay, buckle up: it's going to be a bumpy ride.

P.S. In the following pages, I've tried to be as clear as I can about what is my opinion, what is widely accepted, and what is controversial. Almost every sentence that isn't my opinion is backed up by at least one paper from the literature. I also interviewed more than eighty scientists to make sure I was translating correctly. You can find a full list of every single paper I read and see a list of all the scientists I interviewed at ingredientsthebook.com. Wherever possible, I've linked to the papers I used, so you can read them yourself (or, if they're behind a paywall, you can read a short free summary).

PART 1:

WHY DOES THIS STUFF EVEN EXIST?

"HOW TO DO A COFFEE ENEMA
(BEHIND-THE-SCENES IN MY BATHROOM)"
—Title of a YouTube video

CHAPTER ONE

PROCESSED FOOD IS BAD FOR YOU, RIGHT?

This chapter is about ingredient labels, diabetes, uninhabited islands, porn, and homemade Cheetos.

The road to hell sure isn't paved with butter anymore.

It's cobblestoned with Reese's, studded with Gushers, and sprinkled with Cheeto dust. Your chariot is made entirely from Snickers and Twix, with Oreo wheels, pulled by Haribo horses.

The road to hell is a bunch of industrial, unnatural chemicals made in unholy imitation of food, embalmed in a bright box, and marketed to within an inch of its life. Simply put: processed food is poison.

Right?

Well, it's clearly not a *literal* poison. Eating a Cheeto isn't going to immediately kill you unless it's laced with a gram or two of cyanide. But what if you eat two bags of Cheetos every day for thirty years? That's 21,915 bags—more than 1,300 pounds—of Cheetos. How would that change your risk of a heart attack, or cancer, or death? And how would we *know* Cheetos did the deed? You can't drag a Cheeto into Judge Judy's courtroom. And even if you could, you'd be unlikely to get a conviction without grainy CCTV video of that puffed piece of cheese-coated cornmeal taking a machete to the victim's heart. And you can forget about the other Cheetos in the bag incriminating their friend. Cheetos don't snitch.

Processed food legal proceedings notwithstanding, there *must* be answers to these questions out there somewhere. Processed foods either do or don't increase your risk of cancer. They either do or don't increase your risk of a heart attack. They either are or are not bad for you. If you're thinking: *I already know they're bad for me, because when I eat them, I feel like crap,* I hear ya. I'm all for listening to your body, and that's a valuable data point for your everyday life. But you could just be experiencing a nocebo effect, which is like the placebo effect for bad stuff: if you expect something to feel like crap, it will. Even if it's not all in your head, feeling like crap doesn't give you the kind of information you need to make long-term decisions. There are lots of things that make you feel like crap that *don't* affect your long-term risk of death or disease, like the common cold or calling your cable company. And there are things that feel great that *dramatically* affect your long-term risk of death or disease, like smoking cigarettes.

For long-term decisions, you'd like to know:

1. **How much processed food *exactly* is bad for you?**

2. **Does doubling your Cheeto consumption double your risk? Or do you have to eat a threshold number of Cheetos before anything bad happens?**

3. **How much life does every additional Cheeto suck from your body?**

4. How bad is bad? How many years of life can you expect to give up in return for your processed-food habit?

I thought the answers to these questions just existed out there in the ether, and all I had to do was Google them. Turns out, they do—sort of. And I found them—sort of. But I also found a lot more. What I learned changed the way I look at food . . . but not how I expected. It wasn't like changing your mind from one extreme to the other. I didn't *stop* seeing Satan in the soggy crumbs of milk-soaked Oreos and instead *start* hearing choirs of castrated Chester Cheetahs in angelic harmony. It wasn't like that at all. It was as if another dimension had been added to my existence.

We're going to start exactly where I did: with processed food. In Part I, we're going to worry ourselves sick about it, and we'll talk about why it even exists in the first place. In Part II, we'll look beyond processed food at some of the chemicals we expose ourselves to on a daily basis—from Cheetos to sunscreen to cigarettes. In Part III, we'll come back to all the terrifying numbers you're about to read in this chapter and we'll ask: How does science come up with these numbers? Finally, we'll try to figure out what all this means for you.

Without further ado, let's start at the beginning. To figure out if processed food is bad for you, we have to define processed food. Why? Think of the following (totally hypothetical) experiment to test whether processed food affects blood pressure:

1. **You lock one hundred people in a room.**

2. **You feed half of them a diet chock-full of processed food and the other half no processed food.**

3. **You measure their blood pressure over the next ten days.**

To do this experiment, everyone has to agree on what processed food *is,* because . . . someone has to actually go shopping and buy all the processed food that your human guinea pigs are going to eat.

Sounds pretty obvious, right? But if the definition of "processed food" isn't crystal clear, then the result of the experiment won't be, either. Imagine if the person doing the shopping was told to buy all foods that were sold in a wrapper. Seems pretty straightforward. But that person could buy fancy gold-foil-wrapped pears or Twix, plain oatmeal or Lucky Charms, a freshly baked baguette or Pepperidge Farm Raisin Cinnamon Swirl bread. If the definition of the thing you're testing isn't clear, your results could be all over the map:

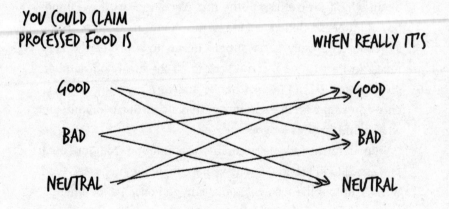

YOU COULD CLAIM PROCESSED FOOD IS

WHEN REALLY IT'S

GOOD — GOOD
BAD — BAD
NEUTRAL — NEUTRAL

In other words, a complete clusterwhoops. So, to scientifically test whether processed food is going to send you to an early grave, we first have to define "processed food."

Okay. Easy—like sorting at Hogwarts, right?

Jam → not processed (GRYFFINDOR!)

Oreos → processed (SLYTHERIN)

Tortillas → GRYFFINDOR!

Cheetos → SLYTHERIN

Olives → GRYFFINDOR!

Starburst → SLYTHERIN

Though it pains me to say it, the Hogwarts method is unscientific. *All* the foods above—whether Gryffindor or Slytherin—are processed in some way. So essentially what we've just done is sorted these foods based on how good or bad we feel about them. But dumping all the foods above into the "processed" bucket doesn't seem right, either. For one thing, "processed" feels like a meaningless category if it's wide enough to hold both "jam" and "Cheeto." For another, the list of "not processed" foods would be way too short—basically raw meat and veggies.

It just *feels* like there should be something fundamentally different about processed and unprocessed foods, kind of like it feels like there should be something fundamentally different about the classic kids' tale *Harry Potter and the Sorcerer's Stone* and the freely-available-on-Pornhub *Hairy Smallballer and the Failure to Bone*,[*] even though both movies are about Harry and Hermione never having sex.

One popular definition of "processed food" is based on how complicated the food seems. This boils down to two things: how many ingredients it has and how pronounceable these ingredients are. Chemists tend to dismiss this definition as nonsensical foolishness, but I think it's worth spending a little time here. First of all, credit where it's due: this definition is clear and simple. But

....
[*] Okay, I made this title up.

it's not great if you're trying to do any science on processed foods. Why? Imagine if you came up with a "processed-food index" (PFI) based on these two metrics. Something like this:

$$PFI = \text{NUMBER OF INGREDIENTS} + \text{NUMBER OF SYLLABLES IN THE NAMES OF ALL THE INGREDIENTS}$$

Here's the PFI for Skittles:

INGREDIENTS	SYLLABLES
SUGAR	2
CORN SYRUP	3
HYDROGENATED PALM KERNEL OIL	9
CITRIC ACID	4
TAPIOCA DEXTRIN	6
MODIFIED CORN STARCH	5
NATURAL AND ARTIFICIAL FLAVORS	10
RED 40 LAKE	4
TITANIUM DIOXIDE	7
RED 40	3
YELLOW 5 LAKE	4
YELLOW 5	3
YELLOW 6 LAKE	4
YELLOW 6	3
BLUE 2 LAKE	3
BLUE 1	2
BLUE 1 LAKE	3
SODIUM CITRATE	5
CARNAUBA WAX	4

PFI = 19 INGREDIENTS + 84 SYLLABLES = 103

Here's the PFI for Smarties:

$$PFI = 9 + 34 = 43$$

Here's the PFI for coffee:

$$PFI = ROUGHLY\ 1,000^{*} + ROUGHLY\ 4,000 = ROUGHLY\ 5,000$$

Intuitively, Skittles and Smarties are probably equally processed, but according to the PFI, Skittles are 2.4 times as processed. Coffee, which is roasted and then extracted with hot water (relatively simple processing) is, according to the PFI, 49 times as processed as Skittles and 116 times as processed as Smarties.

The problem is that the PFI doesn't actually measure processing; it measures how the FDA regulates ingredient labels and how chemists name molecules. For example, enriched flour contains a molecule that goes by three different names:

riboflavin

vitamin B$_2$

**7,8-dimethyl-10-[(2S,3S,4R)-2,3,4,5-
tetrahydroxypentyl]benzo[g]pteridine-2,4-dione**

All three names refer to the identical molecule, but they would yield wildly different PFIs. That problem gets even worse with more complicated mixtures of molecules, like coffee. It doesn't have an ingredient label at all, so what should you use to

.........:.............

* Coffee beans are made of living cells, which are themselves made from thousands of chemicals. More than 950 different chemicals have been identified in roasted coffee, and there are probably many more we haven't detected or identified yet.

calculate the PFI? "Coffee" (PFI = 3), *Coffea arabica* (PFI = 6), or, as I chose to do above, a list of all the chemicals we currently know to be in a cup of coffee (PFI = 5,000)? Depending on what you choose, coffee appears to be either one-thirtieth as processed or 49 times as processed as Skittles.

So some sort of intuitive "ingredient complexity" scale might be fine for quick comparisons in the grocery store, but it won't do for science.

Coming up with a reasonable food processing index that will work in scientific experiments is hard. Carlos Monteiro, a nutritionist and researcher in public health, has, along with his team, proposed a system called the NOVA food classification system. NOVA classifies foods based on the "nature, extent, and purpose" of food processing. In other words: How was the food processed, how much was it processed, and why was it processed? Instead of a numerical scale or two simple buckets—processed or not—the NOVA system has four, ranging from "unprocessed or minimally processed foods" all the way up to "ultra-processed foods." Here are some examples of what might go in each bucket:

GROUP 1: Edible plants, animals, or parts of plants and animals, and any of these when processed to preserve them in their (mostly) original form. Monteiro puts foods like milk, dried fruit, rice, plain yogurt, and coffee in this group.

GROUP 2: Stuff that you would use as an ingredient but not eat by itself. For example, butter, sugar, salt, and maple syrup.

GROUP 3: Foods made by adding group 2 foods to group 1 foods. Ham would fall into this category, as would jams and jellies, canned tuna in oil, and fresh breads.

GROUP 4: Sodas, ice cream, chocolate, instant anything, baby formula, energy drinks, most breakfast cereals, candy, packaged breads, and lots of other stuff . . . including Cheetos.

This seems pretty intuitive, but before we get any further, it's worth noting that the NOVA system is a drastic departure from the way food is currently studied. Most nutrition research today focuses on what's *in* the food. NOVA focuses mostly on what was *done* to the food. The easiest way to see this is by looking at some nutrition facts.

FOOD A

NUTRITION FACTS
SERVING SIZE   100 G
CALORIES   160
TOTAL FAT   14.7 G
TOTAL CARBS   8.5 G
DIETARY FIBER   6.7 G

FOOD B

NUTRITION FACTS
SERVING SIZE   100 G
CALORIES   23
TOTAL FAT   0.4 G
TOTAL CARBS   3.6 G
DIETARY FIBER   2.2 G

From the perspective of what's *in* them, the two foods above are about as different as they could possibly be. Food A has more than twice the carbs, 3 times the fiber, and 37 times as much fat as food B (oh, and lest we forget, 7 times the calories). And yet these foods are both in NOVA group 1. (Food A is avocado, food B is spinach.)

Here's another example:

FOOD C

```
NUTRITION FACTS
SERVING SIZE        100 G
─────────────────────────
CALORIES            304
TOTAL FAT           0 G
TOTAL CARBS         82.4 G
   DIETARY FIBER    0.2 G
```

FOOD D

```
NUTRITION FACTS
SERVING SIZE        100 G
─────────────────────────
CALORIES            375
TOTAL FAT           0.1 G
TOTAL CARBS         93.5 G
   DIETARY FIBER    0.2 G
```

The foods above both have roughly the same number of calories, fiber, sugars, and fats; but food C is NOVA group 2, and food D is NOVA group 4. Guess what each one is.[*]

NOVA's emphasis on what was done *to* a food rather than what's *in* a food is not accidental. It's based on the theory that *what is done to foods* is "the most important factor . . . when considering food, nutrition and public health," as Monteiro puts it. Bold strategy, Cotton, but it appears to be paying off for him: the World Health Organization, the Pan American Health Organization, and the Food and Agriculture Organization of the United Nations have leaned into the NOVA system pretty hard.

Group 4 is the heart of the NOVA system. These foods are what Monteiro calls "ultra-processed foods" or "ultra-processed foods and drinks," and they're defined as "not modified foods but formulations made mostly or entirely from substances derived from foods and additives, with little if any intact Group 1 food." Ultra-processed

....
[*] On the left: honey. On the right: jelly beans.

foods include additives that aren't found in other foods, including flavorings, dyes, and this delicious-sounding list of stuff: "carbonating, firming, bulking and anti-bulking, de-foaming, anti-caking and glazing agents, emulsifiers, sequestrants and humectants." But the definition goes beyond what's added to the food: Ultra-processed foods, according to Monteiro, are produced in industrial processes and designed to be inexpensive and convenient. Finally, they are "packaged attractively and marketed intensively."

You probably haven't heard it phrased quite like this, but this is a description of what you intuitively recognize as "processed food": absurdly cheap, ridiculously convenient, universally delicious, and barely recognizable as food. The NOVA classification is, in essence, a reasonably systematic way of telling whether you're watching *Sorcerer's Stone* or *Failure to Bone*.

Let's review the research done using the NOVA classification.

———

I was *very* surprised when I realized just how much of our diet is made up of ultra-processed foods. In the U.S., over 58 percent of our calories come from ultra-processed foods. More than half! Canada is not much better at 48 percent; France is holier-than-thou (as per usual) at 36 percent. The U.S. is bad compared to France, but we're ahead of Spain (61 percent), and we look like health nuts compared with Germany and the Netherlands at 78 percent! Some of these numbers are so high, they set off my bullshit radar. But on the other hand, these percentages are all based on calorie intake, and ultra-processed foods are usually calorie dense. If, for example, all you ate on a given day was a two-liter bottle of Coke and fourteen cups of raw spinach, you would have gotten 90 percent of your calories from ultra-processed foods that day. Or if you happen to pull through a Sonic drive-through and order the (very tempting) large Oreo peanut butter shake, you would have to eat two whole sticks of butter or 232 cups of spinach to get your ultra-processed food intake down to 51 percent.[*]

So we're clearly eating a *lot* of ultra-processed food. *But is it killing us? And if so, how?* Ultra-processed food could be killing us in a variety of ways: there could be too many chemicals that are toxic to us, there could be not enough chemicals that are good for us, or ultra-processed food could simply be making us obese, and *that* could be killing us.

So one important question is: Does processed food actually make you obese? The hypothesis goes like this: Over the past two hundred years, there has been an incredible increase in the availability of ultra-processed foods, which are extremely calorie dense, cheap, and very convenient. They're also designed to be addictive, so you eat more—specifically, more sugar and fat, and less fiber and micronutrients. Over time, this makes you more overweight or obese, which increases your risk for basically every disease, but especially diabetes, heart disease, and cancer. Transnational food conglomerates don't seem to care, because they're basically just following the tobacco industry's playbook: they're happy to rake in the cash now and kill people later.

We already know some parts of this hypothesis are true. For example, ultra-processed foods as Monteiro defines them were invented extremely recently in human history. Coke, Dr Pepper, Hershey's, Wrigley's, Post cereal, Cracker Jack, Breyer's, Cadbury's, Entenmann's, Pepsi, Jell-O, and Tootsie Rolls were all founded or invented in the thirty-year period between 1877 and 1907. And we are definitely eating more of these foods as time goes on: even if you don't trust the food survey numbers above,

....

[*] For those of you doing the math, that means a large Sonic Oreo peanut butter shake has more calories than two sticks of butter or 232 cups of raw spinach. (Butter is a group 2 food; raw spinach is a group 1 food.)

the fact that every single one of us knows what a Starburst is shows how prevalent ultra-processed foods are. And obesity really *is* getting worse. In the U.S., there are more than twice as many obese people as people who smoke, and this number has been steadily and reliably increasing for years despite the best efforts of every "health" magazine on the planet.

Here's where it's tempting to jump to conclusions. We have before us two propositions: America is getting fatter, and America is eating boatloads more ultra-processed foods than before. Sticking a "because" in between those two propositions would be the easiest thing in the world. But there have been lots of other changes in American society: Office jobs mostly involve sitting on our asses all day; we are more financially and mentally stressed than before; we have invented entirely new ways to feel self-conscious, depressed, and jealous of our friends, thanks to phones and anti-social media; and you can probably think of fourteen other factors in your life that would cause you to down an entire party-size bag of Cheetos in one sitting. Some of the scientists I interviewed suggested that a small part of the obesity epidemic could be explained by people quitting smoking, because nicotine is an appetite suppressant. One even thought that the layouts of houses might play a role: in newer houses, the kitchen—with all its food—is the heart of the home, so it's easier to satisfy food cravings. Oh, and let's not forget our genes. For almost all of human history, food was incredibly scarce, so we evolved to hoard excess calories. Now that excess calories are everywhere, everyone is hoarding them—also known as "getting fat."

It's possible that all these things are equally to blame; it's also possible that ultra-processed foods are the main contributor and all the other factors are just icing on the Cheeto.

If you wanted to try and figure out whether ultra-processed food makes you obese, you might do something like this:

1. Find a huge group of, say, 20,000 people willing to sign over their entire lives to you.

2. Find two uninhabited identical islands about two hundred miles apart and build identical hotels on each.

3. Split the 20,000 people into two groups of 10,000 each and lock each group in its own Hotel California.

4. Feed one group a diet high in ultra-processed foods and the other a diet low in ultra-processed foods for a few decades.

5. Record what happens.

6. Critically, both groups would never be allowed to leave, swim to the other island, or receive food from family or friends back home.

This type of experiment, where you require groups of people to do different things, is called a randomized controlled trial. At the end of the trial, you compare the risk of becoming obese in the high ultra-processed-food group vs. the low ultra-processed-food group. Divide one risk by the other and you get something called the *relative* risk. If you've ever been on the Internet, you've seen relative risks before. This is from an NPR story I found by simply Googling "egg risk": "If you consumed two eggs per day, there was a 27 percent increased risk of developing heart disease, says researcher . . ." (Don't worry, we'll come back to whether you should eat eggs.)

Most relative risks that have to do with food, including the egg risk, do *not* come from randomized controlled trials. Instead they come from experiments in which you recruit a bunch of people and check in with them regularly for years, but you don't require them to change their diet or behavior in any way. This type of

experiment is called a prospective cohort study. At the end of o.
of these studies, you categorize people by how much ultra-
processed food they ate. Then, just like in a randomized controlled trial, you compare the risk of becoming obese in the low
ultra-processed-food category vs. the high ultra-processed-food
category. Divide those two numbers and you get the relative risk.

A relative risk has the same meaning whether it comes from a
randomized controlled trial or a prospective cohort study: it gives
you an indication of how screwed you are relative to someone
else. If your neighbor's risk of getting mauled by a mountain lion
is 25 percent and yours is 40 percent, then your relative risk vs.
your neighbor's is 40/25 = 1.6, which means:

You're 1.6 times as screwed as your neighbor.

You're 160 percent as screwed as your neighbor.

You're 60 percent more screwed than your neighbor.

These are three ways of saying the same thing: when it comes
to mountain lions, it's better to be your neighbor than you. Most
relative risks have nothing to do with mountain lions and everything to do with your health. Let's look at a few, specifically some
that deal with ultra-processed food.

Carlos Monteiro's NOVA classification system is fairly new, so
there aren't that many studies that use it. There has been only one
prospective study testing the link between ultra-processed foods
and obesity: an 8,000-person trial in Spain with about nine years
of follow-up. The authors found that people who ate about 4 times
as much ultra-processed food had a 26 percent higher risk of becoming overweight or obese over the course of nine years.

What about other outcomes?

Researchers in France recruited over 100,000 people and followed them for an average of five years, diagnosing cases of cancer as they came up. They found that people who consumed, on
average, about 4 times as much ultra-processed food had a
roughly 23 percent higher risk of developing any cancer. Using

he data set of French folk, another set of researchers found ople who ate more than twice as much ultra-processed l a roughly 25 percent higher risk of developing irritable ndrome. Going back to the data set in Spain, researchers at people who ate more than 2.5 times as much ultra-processed food had a roughly 21 percent higher risk of developing high blood pressure over nine years. And now, for the putrid cherry on this sundae of despair: using the French data set, the same group of researchers who found a higher risk of IBS also found that people who ate 10 percent more ultra-processed food had a 14 percent higher risk of death.

I will readily admit that I was somewhat surprised by these results. Not gonna lie: I kinda freaked out a little bit. A 23 percent higher risk of developing cancer? A 25 percent higher risk of IBS? A 26 percent higher risk of becoming obese? A 14 percent higher risk of death? *How is this crap even legal!?*

Okay, I freaked out a lot.

I freaked out for two reasons: First, those numbers are legit terrifying; second, I was trained as a chemist.

That second thing seems like a not very good reason to be freaked out, but let me explain. Imagine that you have two balloons in front of you, each full of pure cyanide gas. One balloon contains cyanide harvested from hand-selected seeds from apples growing naturally in organic orchards in Massachusetts. (Yes, there's cyanide in apple seeds. More on that soon.) The other balloon contains cyanide produced in the Andrussow process, in which methane and ammonia are burned in oxygen at over 2,000°F (1,093°C) in the presence of platinum. Which balloon is safer to inhale?

Neither, of course. Both will kill you. To a chemist, this is an axiom of biblical proportions: if two molecules have the same chemical structure, they will do the same thing to your body. Cyanide produced by apples and cyanide produced by man are both *cyanide*. Now, switch out the word "cyanide" for "pound cake" and you have an axiom that is slightly less biblical but still makes perfect sense to a chemist: a pound cake baked in Ina Garten's kitchen and a pound cake produced in a factory are both pound cakes. So the notion that they would have wildly different effects on your health just doesn't feel right to a chemist, even if the factory version has a few additives. But this is exactly Monteiro's argument: that what is done *to* a food is more important than what the food *is*. To a chemist, this is like saying "Natural cyanide is less toxic than industrial cyanide," which of course makes no sense. But to most nonscientists, Monteiro's argument is powerful, intuitive, and obvious. And this difference in perspective almost always produces the same conversation. Whenever a hard-core chemist and a regular human talk about food, the result is something like this:

"THE CONVERSATION" FROM A CHEMIST'S POINT OF VIEW

HIPPIE

I buy only organic, all natural, raw, unprocessed foods.

CHEMIST

Those words don't mean anything.

HIPPIE

Yes they do! It means my food isn't loaded up with chemicals.

CHEMIST

That's not really a reasonable indictment of foods, since every food is chemical. In fact, did you know that everything on planet Earth is made of chemicals, including you?

HIPPIE

My body is a temple.

CHEMIST

Your body is a large, mostly empty space into which only priests are allowed?

HIPPIE

I just think natural food is healthier, that's all.

CHEMIST

(Face-palms so hard that nose breaks.)

And here it is again, this time from the regular human's point of view:

"THE CONVERSATION" FROM A NON-CHEMIST'S POINT OF VIEW

CONCERNED CONSUMER

I want to choose healthy foods, but it's hard to do all the research and know whom to trust. So I buy organic and natural stuff, 'cause that feels better, and maybe it's a bit healthier.

PRO-GMO PHARMA BRO

You're falling for a dumb marketing trick.

CONCERNED CONSUMER

But what about all the chemicals they're adding to foods? I have no idea what that stuff is . . .

PRO-GMO PHARMA BRO

ALL FOOD IS MADE OF LITERALLY NOTHING BUT CHEMICALS. YOU ARE 100 PERCENT CHEMICALS. THE WORLD AROUND YOU IS 100 PERCENT CHEMICALS. EVERYTHING THAT EVER WAS OR WILL BE IS A CHEMICAL!

CONCERNED CONSUMER

There's really no need to yell.

PRO-GMO PHARMA BRO

DO NOT QUESTION ME, PEASANT.

CONCERNED CONSUMER

Just let me buy my all-natural, organic, no additives, no hormones, unprocessed groceries. Oh, and fuck off.

PRO-GMO PHARMA BRO
(Punches self in face for no apparent reason.)

Here's the conversation again, this time boiled down to its essence:

Hippie: Chemicals are bad.

Pharma Bro: Everything is chemicals.

Both of these arguments are absurd.

To the hippie, I would say: *Really, all chemicals are bad? Including water, air, and all food?*

To the pharma bro, I would say: *Do you not speak English?* The hippie *obviously* means *Chemicals on ingredient labels that I can't pronounce and don't recognize are bad.* So instead of picking a pedantic fight over the *literal* definition of the word "chemical," try responding to the hippie's *actual* concern: some chemicals are bad for your health, and it's hard to know which ones.

I used to be solidly in the "Everything is a chemical, doofus" camp. But when I read the studies claiming double-digit percentage increases in risks of getting a variety of diseases *caused by the way a food was processed*, for the first time in my life I thought: *Hell, maybe the hippies are right.* This research was a direct challenge to everything I thought I knew. Could ultra-processed bread sold in plastic bags really and truly be worse for you than the bread baked in the store? What about those frozen lemonade cylinders: Were those really and truly worse for you than squeezing a couple lemons into three cups of sugar? And what about Cheetos?

The ones you buy in the store are made by forcing cornmeal through a corkscrew that generates so much friction that it boils the water in the cornmeal, which, as it expands, creates a random assortment of air pockets and the characteristic Cheeto puff shape. But as alien as this process seems, you can make pretty decent imitation Cheetos from the convenience of your very own kitchen. As part of the research for this book, I talked with food historian Ken Albala, who had coincidentally made fake Cheetos the day before. His improvised recipe went like this:

1. **Cook some rice noodles.**

2. **Dry them in a dehydrator (basically a super-low-temperature oven with air vents designed to evaporate most of the water out of food).**

3. **Spray the dry noodles with oil spray.**

4. **Microwave them until they puff up.**

5. **Sprinkle with spicy powder of your choice, e.g., sriracha.**

6. *Et voilà!* **Sorta-Kinda-Flamin'-Hot Cheetos.**

The good folks over at *Bon Appétit* developed a *very* involved recipe that more faithfully replicates the Cheeto experience; you can watch it on the interwebs. Whether you eat Ken Albala's improvised Cheetos or Claire Saffitz's gourmet Cheetos, or just go buy a bag of real Cheetos, you're still eating a bunch of carbs with some flavors and spices. So my gut reaction as a chemist to the idea that factory-made Cheetos are worse for you than homemade/natural/organic Cheetos is: No.

And yet that idea is exactly what my quick scan of the data seemed to suggest: people who ate more ultra-processed food were in poorer health and had a higher risk of death.

Well, damn.

So . . . who's right?

―――――

Before we dive in, let's pause and acknowledge that you're not *just* interested in the health effects of ultra-processed foods. You care about *all the foods*! *Should I eat processed food?* is just the tip of the question iceberg. The real question is: *What should I eat?*

Before we get to THE ANSWER, know this: whatever THE ANSWER may be, there are a small number of very loud people who passionately believe that they know it. (Does "Eat food, not too much, mostly plants" sound familiar?) Depending on the food (or sunscreen, or makeup, or cleaning product), there can be almost universal agreement or visceral, holy-war levels of disagreement. Nowhere is this more prominently displayed than in what we call a "diet." There are a ridiculous number of diets available to try, and new ones pop up all the time. But when you take out all the window dressing, diets are really just two lists: a good list and a bad list; a list of foods you should eat, and a list of foods you shouldn't. That's it. This deceptively simple two-list structure nevertheless results in an overwhelming variety of programs to choose from. Recent diet books include:

The Paleo Diet

The Flex Diet

The Simple Diet

The 3-Season Diet

The Easy-Does-It Diet

The Aquavore Diet

The Peanut Butter Diet

The Supermarket Diet

The Good Fat Diet

The Belly Melt Diet

The 5-Bite Diet

The Dakota Diet

The Scripture Diet

The Uncle Sam Diet

The Plateau-proof Diet

The 4 Day Diet

The 17 Day Diet

The Alternate-Day Diet

The 20/20 Diet

The No-Time-to-Lose Diet

The Thermogenic Diet

The G.I. Diet

The Good Mood Diet

The Salt Solution Diet

The Nordic Diet

The Thin Commandments Diet

The Great American Detox Diet

The Better Sex Diet

The Sleep Diet

The Couch Potato Diet

The Self-Compassion Diet

The No S Diet

The Lemon Juice Diet

The Baby Fat Diet

The Yoga Body Diet

The Four-Star Diet

The Warrior Diet

No-Fad Diet*

The Martini Diet

and of course

The Diet to End All Diets: Losing Weight God's Way

Diet books are like names of British pubs: mostly random and nonsensical, but, man, do they sound good! And the similarities don't end there. There are some diet books that are just as old as British pubs. For example, here are two books, one published in 1870 and the other published in 2018. Guess which is which:

Book A:

THE DISCOVERIES AND UNPARALLELED EXPERIENCE OF PROF. R. LEONIDAS HAMILTON, M.D., WITH REGARD TO THE NATURE AND TREATMENT OF

....
* So ironic!

Diseases of the Liver, Lungs, Blood, AND Other Chronic Diseases; CONTAINING, ALSO, A Biographical Sketch of His Life, (From Harper's Magazine): WITH HIS Common Sense Theory of Diseases AND THE Evidence of his Wonderful Cures.

BOOK B:

Medical Medium Liver Rescue: Answers to Eczema, Psoriasis, Diabetes, Strep, Acne, Gout, Bloating, Gallstones, Adrenal Stress, Fatigue, Fatty Liver, Weight Issues, SIBO & Autoimmune Disease.

Book A is the old one; all those CAPITAL LETTERS give it away. And by the way, our modern obsession with "natural" is not modern. Here's a book title from 1889: *THE PERFECT WAY IN DIET: A TREATISE ADVOCATING A RETURN TO THE NATURAL AND ANCIENT FOOD OF OUR RACE.*

Yikes.

Wine drinkers will enjoy this classic from 1724: *The JUICE of the GRAPE: or, WINE preferable to WATER. A TREATISE, WHEREIN WINE is shewn to be the Grand Preserver of Health, and Restorer in Most DISEASES. With many Instances of Cures perform'd by this Noble Remedy; and the Method of using it, as well for Prevention as Cure. With a Word of Advice to the Vintners.*

Lest teetotalers feel left out, we also have, from 1779, *A TREATISE ON THE USE AND ABUSE OF MINERAL WATERS. WITH RULES for DRINKING the WATERS; AND A PLAN of DIET for IN-VALIDS LABOURING UNDER CHRONIC COMPLAINTS.*

In 1916, Eugene Christian wrote a five-volume (!) diet book: *ENCYCLOPEDIA OF DIET: A TREATISE ON THE FOOD QUES-TION IN FIVE VOLUMES EXPLAINING, IN PLAIN LANGUAGE, THE CHEMISTRY OF FOOD AND THE CHEMISTRY OF THE HUMAN BODY, TOGETHER WITH THE ART OF UNITING THESE*

TWO BRANCHES OF SCIENCE IN THE PROCESS OF EATING SO AS TO ESTABLISH NORMAL DIGESTION AND ASSIMILATION OF FOOD AND NORMAL ELIMINATION OF WASTE, THEREBY REMOVING THE CAUSES OF STOMACH, INTESTINAL, AND ALL OTHER DIGESTIVE DISORDERS.

Dieting and health is one of the most resilient book genres of all time. Once Gutenberg finished printing the first run of the Bible, he started on diet books and hasn't stopped since. And of course the fun is not limited to books. There is also the Internet, where people will tell you to empty a turkey baster full of coffee into your butt (room temperature, please!) or drink your own pee. When it comes to food and health, there is an overwhelming tsunami of information that started at least three hundred years ago and hasn't stopped since. In short: Googling diets or health usually leaves you confused, worried, and 7 percent more likely to stick pus up your nose.

So, on the one hand, the numbers on ultra-processed food are pretty scary. On the other hand, we've been glorifying and vilifying different classes and categories of food for hundreds of years. Who's to say this whole ultra-processed food thing isn't the latest fad? On the third hand, it seems powerfully intuitive that the further a food is from nature, the worse it is for you.

Over the rest of Part I, we're going to take a wandering but ultimately purposeful path toward clarity. First we'll look at the chemical reaction that generates every ounce of food on this planet. Then we'll consider three important reasons our ancestors processed food.

Reason 1: to avoid an immediate and painful death.

Reason 2: to avoid a slower but no less painful death.

Reason 3: just for fun.

But before we get to death and fun, let's start where all food starts: with the most important chemical reaction on earth.

CHAPTER TWO

PLANTS ARE TRYING TO KILL YOU

This chapter is about carbon dioxide, pooping, plumbing, the Energizer Bunny, condoms, poisonous potatoes, and NASA ice cream.

Long before humans started taking down big game and cooking the kill with fire, we relied mostly on plants for food. And we weren't alone: every animal on Earth either ate plants or animals that ate plants; or ate animals that ate animals that ate plants; or ate animals that ate animals that ate animals that ate plants; or ate animals that . . .

You get the picture.

Plants are basically magic: they build themselves out of air and soil using energy from the sun. They literally feed the whole planet, either directly or indirectly. What's their secret? You've heard the answer before; you probably learned it in high school: photosynthesis. And you may have seen the chemical reaction:

$$6CO_2 \text{ (GAS)} + 6H_2O \text{ (LIQUID)} \rightarrow C_6H_{12}O_6 \text{ (IN SOLUTION)} + 6O_2 \text{ (GAS)}$$

Or, if pictures are your jam:

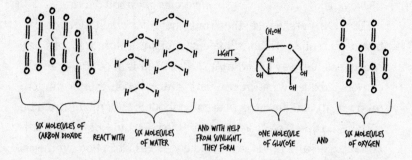

SIX MOLECULES OF CARBON DIOXIDE REACT WITH SIX MOLECULES OF WATER AND WITH HELP FROM SUNLIGHT, THEY FORM ONE MOLECULE OF GLUCOSE AND SIX MOLECULES OF OXYGEN

(By the way, if you've ever googled a chemical, you've probably come across structures like the ones above. They're a chemical shorthand. Each letter corresponds to an atom. C = carbon; O = oxygen; H = hydrogen. Lines represent chemical bonds: in this case, electrons shared between atoms. Wherever two or more lines intersect, there's a carbon atom; it's not drawn explicitly, but it's there. Why don't chemists just draw out every single carbon? In large molecules, it would take forever.)

Here's the explanation of photosynthesis you got in high school:

Plants use light energy from the sun to convert six molecules of carbon dioxide and six molecules of water into a molecule of glucose and six molecules of oxygen.

I fell asleep so fast my head hit the desk hard enough to wake me up again. Let's break this down:

Plants use light energy from the sun . . .

Humans invented solar panels in the 1950s. Plants invented them *500 million years ago*. Because, yes, plant leaves[*] are basically mini solar panels. Plants figured out a way to build tiny molecular machines that change their shape and behavior in

....

[*] And any other photosynthetic-capable cell, the most important example of which is algae in the oceans.

response to being hit by a single photon of light, and harness that energy to make sugar.

Okay, next: . . . *to convert six molecules of carbon dioxide* . . .

From our perspective, the atmosphere has way too much carbon dioxide in it. (Hello, climate change!) But from a plant's perspective, it has way too *little*. At sea level, air is about 0.04 percent carbon dioxide. What this means is that if you took a random group of 10,000 air molecules, only about four of them would be carbon dioxide. The rest—9,996 of them—would be . . . *not* carbon dioxide, and therefore absolutely useless for photosynthesis. So plants have somehow managed to extract—from this veritable ocean of mostly useless crap—the four molecules out of every 10,000 that they need.

Let's keep going: . . . *and six molecules of water* . . .

We all need that sweet, sweet water.

Almost there: . . . *into a molecule of glucose* . . .

The glucose plants make is used in all kinds of ways. It's burned for energy in the same way we humans burn glucose for energy; it's made into sucrose (exactly the same as the sugar in your pantry); it's turned into starch and stored for the winter; it's turned into cellulose, which is used to build the plant . . . and the list goes on. It's basically the Swiss Army knife of the plant world.

And last but not least: . . . *and six molecules of oxygen.*

For every molecule of glucose the plants make, *six* molecules of oxygen are formed. The plant then has to shove those oxygens out into the atmosphere, a place where 2,096 out of every 10,000 molecules are *already* oxygen. Some oxygen is used to break down sugars for energy, but most of it is ejected into the atmosphere. Oxygen is basically photosynthesis exhaust.

All in all, plants use the sun's energy and water to break apart molecules of carbon dioxide and string the carbon atoms together, thus forming chemically stable, water-soluble, ring-shaped energy storage molecules. You know these rings as "sugar." Sugar can be burned for energy immediately, used as a building material, or linked together in chains thousands of units long, to be stored for later use.

Sugar is produced in the leaves, but because it's so important, every other part of the plant also needs it. So it has to travel from the leaves to lots of other places in the plant. In the herbs that are growing in your kitchen, that trip may be only a few inches long. In the tallest trees, it could be hundreds of feet. So how does sugar get from one end of the plant to the other?

Before we get to the *how*, we have to talk about the *how much*. The simple answer is: a lot. An oak tree can make 25 kilograms of glucose every single day. That's the weight of a small child or a female golden retriever. Much of this sugar is transported elsewhere: flowers, fruits, stems, stalks, trunk, and roots.

We humans have a pretty cool circulatory system: we have one powerful central pump (the heart) that pushes a thick liquid dense with living cells (blood) through big arteries, medium-size arteries, and tiny capillaries. Plants have none of that. And yet even the tallest tree in the world, Hyperion in California, somehow manages to move sugar from the tip of its tallest leaf 380 vertical feet above the ground to the tip of its farthest root, which could be as far as 100 feet from the trunk. How? Phloem. You probably learned about phloem in school:

The xylem transports water from the roots up to the rest of the plant and the phloem transports sugar from the leaves to everywhere else.

Phloem is a complicated tissue, but the key components are called sieve tubes. Sieve tubes are basically pipes. Except they're not made of copper like the artisanally crafted taps in Pinterest bathrooms; they're made of living cells. *Single* living cells, laid end-to-end like sections of an oil pipeline; the joints are shot through with holes, like the sieve in your kitchen. Each section, called a sieve element, is only a few hundred millionths of a meter long. In leaves, sieve elements are roughly 10 millionths of a meter wide.* Imagine the amount of sucking (or blowing) power required to pull (or push) a sugar solution through a straw that's only 10 millionths of a meter wide but hundreds of feet long. And yet plants do it every day of the week. How?

Photosynthesis. Unlike you and me, photosynthesis is incredibly productive. Under optimal conditions, some plants can photosynthesize a molecule of glucose using just 60 photons of light. (For reference, about 300,000,000,000,000 photons hit your eye every second when you look up at an empty patch of blue sky on a sunny day.) Even under average conditions, a plant can make about 800 milligrams of sugar per medium-size leaf per day. This sugar is continuously loaded into the sieve tubes in the leaves, and as you might know, the more stuff you try to cram into a limited space, the more pressure builds up in that space. Luckily that sugar has somewhere to go to relieve the pressure: the rest of the plant. Except the pressure is never really relieved, because photosynthesis keeps happening in the leaves, and more and more sugar gets made there, which pushes it into the sieve tube, which pushes the rest of the contents of the sieve tube out to the rest of the plant.†

So you can think of photosynthesis as kind of like a pump, but not a mechanical pump that works by compressing things—a *chemical* pump that works by making more and more sugar until it just has to *go* somewhere, and that somewhere is out the end of the sieve tube.

....

* This is roughly one-tenth the width of a human hair. It's also six-millionths the width of a 1978 Ford Pinto and three hundred-billionths the width of the state of Nebraska.
† This is a slight simplification; in reality, the higher sugar concentration in the leaf *pulls* water into the sieve tube by osmosis, and that's what generates the pressure that pushes the sugar water out of the leaves to the rest of the plant.

But don't let the fact that this pump system isn't mechanical fool you. The pressures generated are . . . well, when you go to the doctor and they measure your blood pressure, if you're healthy (and somewhat lucky), it'll be in the range of 2 pounds per square inch (psi). A car tire is pressurized at about 32 psi, or roughly 15 times higher than your blood. Plants—which, remember, have no centralized pumps—can pressurize their sieve tubes to about 145 psi! To feel what that kind of pressure would be like, you would have to throw on some scuba gear and dive *100 meters below sea level.* The force per square inch on your skin resulting from 328 feet (and 1 inch) of water *pressing down on you from above* is the same as the force per square inch within a teensy-tiny tube one-tenth the width of a human hair nestled deep within the plants that surround you every single day.

So the next time you look at a tree—or even the herbs you forgot to water in your kitchen—take a moment to recognize that you are looking at the most technologically advanced plumbing system on Earth.

Now let's talk about *what* is flowing through this plumbing system. Remember that photosynthesis produces a lot of sugar in the plant leaves. But the plant isn't making *solid* sugar. Almost everything that happens in a plant, including photosynthesis, happens in water. So when plants make sugar, they're making it in water. And when plants transport sugar through the phloem, they're also transporting it in water.

If you dissolve two teaspoons of sugar in a cup of tea or coffee, you're making roughly a 3.3 percent sugar solution. To most of us, this tastes pretty sweet. A can of Coke is about a 10 percent sugar solution.[*] Plant sap *starts* at 10 percent and goes up to 50 percent. So, in some plants, the liquid that is flowing through

....

[*] If you've ever tried to drink a 10 percent sugar/water mixture, hats off. It's pretty disgusting. Coke, orange juice, and other juices have flavorings and a bunch of acid, which essentially disguise the sugar.

their plumbing system has 3 times the concentration of sugar of a can of Coke. Plants are the first and most ancient of syrup producers in the entire world.

So to recap: fruits are delicious and great and all, but the liquid flowing down from the leaves to the roots through the thousands of 10-millionths-of-a-meter-wide tubes at pressures equivalent to those 100 meters below sea level or in an operating firehose—that's the real *mamahamajabanahoooooeeey* of sugar.

And sugar is just the beginning.

If you live in the U.S. or another wealthy country, your food choices are basically infinite. But if you trace those infinite choices back to their source, you'll find just one: plants. Photosynthesis takes two molecules made from three chemical elements—carbon, hydrogen, and oxygen—and turns them into sugars. Plants burn those sugars for energy immediately, but they also store them for later as starch or fat. So three of our most important food groups—sugars, starches, and fats—are all made from the same three elements harnessed by photosynthesis. (Fiber is also made from those same elements; while it's not exactly food, it's extremely helpful for a pleasant pooping experience.)

Plants also make protein. To do that, they need nitrogen. Some plants suck nitrogen up from the soil via their roots. Other plants partner with microbes that can pull nitrogen gas (N_2) out of the atmosphere and make ammonia (NH_3), which the plants can then incorporate into themselves as proteins, vitamins, and DNA.

So, to recap: photosynthesis powers the conversion of carbon, hydrogen, oxygen, and nitrogen into sugars, starch, fiber, fat, and protein. Plants also take up minerals from the soil and make some of the vitamins we need to survive.

In short, plants turn things that are not food into food.

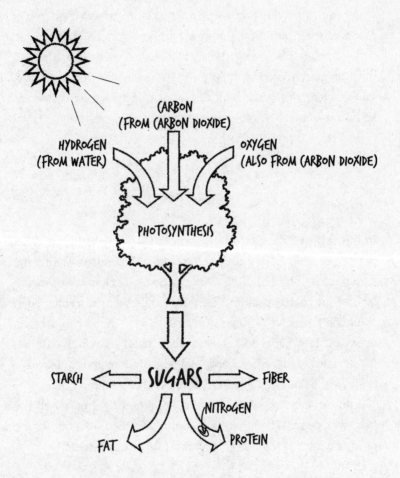

Where do they store all this food? They build themselves out of it. Oh, and let's not forget that plants are mostly water. They have everything you—and other animals—need to live.*

If you're a plant, it's pretty amazing that you can survive off water, air, sunlight, and soil. But there are drawbacks. You're made of food, and you pump a highly nutritious sugar syrup through your plant veins basically 24/7. Also, you're a sitting duck, literally rooted to the ground. Not only can you not move, you also can't growl or bark or bite or punch. For all these reasons, lots of insects and animals would love to eat you.

How do you prevent that?

You have to fight dirty.

———

In the early 1980s, western Victoria, Australia, was in the midst of one of the worst droughts of the twentieth century. Among the sufferers were a herd of fifty Angora goats. Lack of water meant lack of grazeable pasture; the poor goats were starving. Then someone cut down a sugar gum tree. Sugar gums can grow to over 100 feet high, and they're often used as windbreaks on farms. The felled tree probably had tens of thousands of leaves—not a goat's preferred food, but better than nothing, right?

Unfortunately, no. Within twenty-four hours, almost half the herd was dead. (The other half would have died, too, but for the prompt action of the goats' caretaker.) What happened?

Cyanide.

Cyanide is a beautiful molecule.

Fourteen units of negative electrical charge surround, cloud-

....

* Granted, if you eat only one species of plant, you may not get all the amino acids, vitamins, or minerals you need. But if you eat the right mix of varieties, it is possible to get all the required nutrients, even as a strict vegan.

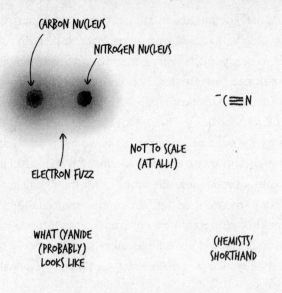

CARBON NUCLEUS

NITROGEN NUCLEUS

$^-(\equiv N$

ELECTRON FUZZ

NOT TO SCALE
(AT ALL!)

WHAT CYANIDE
(PROBABLY)
LOOKS LIKE

(CHEMISTS'
SHORTHAND

like, two small clusters of positive charge, one with six units and the other seven. You can't see the inner positive charges, but the outer, negative layers look like a lopsided cloudy dumbboll, one weight slightly heavier than the other. Negative charge is dense close to the clusters of positive charge but dissipates as you go farther away, like the smell after a fart.

Cyanide is simple. It's made of only two atoms: one carbon and one nitrogen. Cyanide is light: of the molecules we're likely to encounter throughout our travels on this earth, only a handful are lighter.[*] And cyanide is extremely toxic. I weigh about 160 pounds; one-tenth of a gram of cyanide would probably kill me. Half a gram—about the weight of a single regular-size metal paper clip—definitely would. Depending on the dose, I could die quickly, less than sixty seconds after the first molecule passes my lips, though my heart might continue to beat for three to four minutes after my last breath.

Cyanide is so toxic because it *looks* like oxygen but doesn't *act*

....
* These include hydrogen, methane, ammonia, and water.

like it. When you breathe in air, red blood cells absorb oxygen in the tiny little hallways of your lungs. Your blood then carries that oxygen to pretty much every cell in your body, where little mini-cell-like things called mitochondria use it to produce a molecule called adenosine triphosphate, or ATP. You can think of ATP as a molecular AAA battery. These AAA batteries are the main source of energy for most cells in your body, so most cells have lots of mitochondria in them. Oxygen is critical in the very last step of AAA battery production. Electrons (from the chemical bonds in your food) are smashed into an oxygen molecule (from the air you breathe), along with two hydrogen ions (probably from the water you drink), to form a molecule of water and simultaneously power the reaction that makes that AAA battery. Basically:

$$\text{ELECTRONS + OXYGEN + HYDROGEN} \rightarrow \text{WATER + ENERGY TO MAKE AAA BATTERIES}$$

And it's not entirely wrong to say that the above equation can be simplified to:

$$\text{FOOD + AIR + WATER} \rightarrow \text{ENERGY}$$

This reaction is the basis of your life. You eat for the electrons; you breathe for the oxygen; you drink for the hydrogen. Take any one of these away and you die.

Oxygen, electrons, and hydrogen ions need to be arranged perfectly in space throughout a bunch of individual steps for this overall reaction to produce AAA batteries, and your body uses a series of enzymes[*] to make this happen. This is where cyanide

....

[*] An enzyme is a protein that helps a chemical reaction go faster. Enzymes are typically much larger than the molecules involved in the reaction itself.

comes in. Cyanide can pass itself off as an oxygen molecule to one of the critical enzymes in the pathway. When you're not breathing in cyanide, oxygen binds to the enzyme and gets split apart, from one oxygen molecule into two oxygen atoms. If you breathe in cyanide, it quickly diffuses to your mitochondria and binds to the enzyme in oxygen's place. Unlike oxygen, the cyanide molecule doesn't split apart, so the enzyme just sits there, incapacitated. Eventually, the cyanide unbinds, letting the enzyme get back to work, but for the period when it was bound, AAA batteries don't get produced.

ELECTRONS + CYANIDE + HYDROGEN → NOTHING

Your body has thousands of trillions of mitochondria, so if you inhale one molecule of cyanide, it may slightly reduce the number of AAA batteries produced inside one mitochondrion inside one of the roughly 37 trillion cells in your body, but it ain't gonna kill ya. Eventually your body will attach a sulfur atom to the cyanide, forming less toxic thiocyanate, which you'll pee out and then move on with your life. But if you were to inhale enough cyanide, you could prevent a *lot* of mitochondria from making AAA batteries. And what's the Energizer Bunny without the battery in his thigh?

Dead.

High doses of cyanide gas can give you a dry, burning sensation in your throat. Then, no matter how it got into your system, you'd feel like you were suffocating, and you might gasp for air. Then you would stop breathing. Convulsions would follow. Then—mercifully—you'd black out. At this point you might have a heart attack and die relatively quickly thereafter; or, if your brain can manage to keep your heart pumping, it might take a few more minutes before your heart muscles finally run out of batteries. This is more or less the same series of unfortunate events that happens when you can't get enough oxygen into your lungs—although in this case, incredibly, there is plenty of oxygen in your lungs and throughout your body. You just can't use it, because cyanide is in the way, blocking the oxygen from being used. Cyanide is individually suffocating cells in your body—even though there's plenty of oxygen around. It's like dying of thirst in a pool.

If you are a creature that has mitochondria, consuming enough cyanide will probably kill you. And mitochondria are not rare. Angora goats have them. Regular goats have them. You have them. Your dog/cat/gerbil/ferret/lemur/parakeet/mole has them. Insects have them. Mammals have them. Basically, any creature with the desire and capacity to eat plants has mitochondria. If you're a plant and you can make cyanide, the mitochondria of your potential predators provides a very useful target for your poison.

Cyanide is simple: it's made of carbon and nitrogen, both of which plants can get in almost unlimited quantities from the air and the soil. Cyanide is light: it's just two atoms, which means it's energetically pretty cheap to produce compared to, for example, a protein, which has thousands of atoms (or more). And cyanide strikes at one of the core functions of staying alive—generating energy—so it's extremely toxic to a wide range of potential predators. It's the perfect poison . . .

Except for one tiny little detail: plants have mitochondria, too. So cyanide is just as toxic to the plant as it is to a potential predator. There's a way around this, and it's both simple and devious. Instead of making pure cyanide, the plants attach cyanide to a harmless sugar molecule, making what's called a cyanogenic glycoside.

You can think of cyanogenic glycosides as grenades. The explosive part of the grenade is the cyanide molecule, and the pin is the sugar.

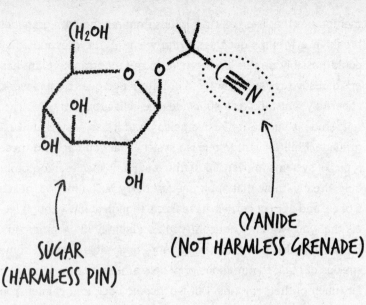

(H₂OH

OH

OH

OH

OH

SUGAR
(HARMLESS PIN)

CYANIDE
(NOT HARMLESS GRENADE)

Pin in: harmless. Pin out: not harmless.

Pulling the pin from a cyanogenic glycoside requires a specific enzyme called beta-glucosidase. You can call this enzyme "Phillip," because it's easier to remember than "beta-glucosidase." For reasons known only to himself and God, Phillip loves pulling pins out of grenades. It's his calling. His nature. His destiny:

PHILLIP + GRENADE → *BOOM!*

Neither grenades nor Phillips are toxic on their own, but together they release cyanide. If a plant were to store these two in the same part of the cell, they would immediately mix, produce cyanide, and severely damage or kill the plant. Not good. So plants store grenades and Phillips separately. During normal plant operation, everything is fine; ne'er the twain shall meet. But if, say, a beetle or a caterpillar comes along and starts eating the plant—ripping, tearing, crushing, masticating the leaves—the membranes that keep all the Phillips from mixing with grenades break. The Phillips get their dying wish: all the grenades they could possibly de-pin. Somewhere in the unfortunate plant eater's digestive system, cyanide is produced and goes right to work, cheerfully whistling as it suffocates the cells around it.

Cyanide is such an effective poison and it's so easy to make a grenade/Phillip cyanide release system that you can find these types of systems in over 2,500 different plant species.[*] You probably already know that apple seeds, cherry pits, almonds, peach stones, and apricot kernels have them, though at low enough levels that you wouldn't notice if you accidentally ate a pit or two. But they're also present at much more noticeable levels in some species of plant that millions of people around the world rely on for much of their calories. But we'll come back to this later. You think cyanide is the only plant poison in town? Nope. Plants are just warming up.

———————————

There are more types of plant poison than there are United States senators, and each general type might have twenty, fifty, or a hundred different specific poisons.

Some are even more devious than cyanide—for example, tannins. Tannins are relatively large molecules, made of tens, hundreds, or even thousands of atoms (rather than cyanide's two), and they work very differently. Instead of preventing mitochondria from using oxygen, tannins stick to proteins. Imagine you're walking from one room in your house to another when two of your young kids grab onto your hands and refuse to move. You could still walk, but you'd have to drag the kids along. Then two more

....

[*] And the cyanide party isn't limited to plants. *Pseudomonas aeruginosa*, a common drug-resistant bacteria, can make and release cyanide when it infects you.

grab you by the legs. At this point it's a bit like wading through molasses. Then another kid latches onto your waist, and two more hang off your neck and shoulders. Eventually, so many kids attach themselves to you that you're paralyzed and totally unrecognizable: covered in kids. This is what tannins do to proteins.[*]

One consequence of eating foods that have a lot of tannins—for example, certain acorns—is that the tannins bind to the proteins in the food, preventing them from being digested. So, unfortunate tannin-eating mammals just poop out those valuable proteins. Chickens fed about a 1 percent tannin diet grow more slowly and make fewer eggs than chickens fed tannin-free diets because they're not getting the full benefit of the proteins they're eating. At much higher doses, tannins become acutely toxic: instead of starving victims, they'll produce ulcers and other intestinal damage. Chickens fed a 5 to 7 percent tannin diet die; others animals, like cows, are more resistant: it takes tannin levels of 20 percent or higher to kill.[†]

The most famous witches' brew in history—courtesy of Shakespeare—includes the root of poison hemlock, which is a veritable . . . witches' brew . . . of chemicals called alkaloids. The caffeine in your coffee is an alkaloid; so is the morphine in your IV drip and the quinine in your gin and tonic. Nicotine, cocaine, and strychnine are all alkaloids, too. In high doses, they can shut down your nervous or respiratory systems. In low doses, some alkaloids can be extremely useful drugs. Before we started making

....

[*] This is why your mouth puckers when you imbibe red wine or other foods with low levels of tannins. The tannins bind to proteins lining the inside of your cheeks, creating that mouth-puckering feeling known as astringency.

[†] Most animals avoid tannins, not so much because of the ulcers or death, but for a much simpler reason: they taste bad. So it turns out that this particular plant poison is actually a feeding deterrent way before it becomes toxic. And in low doses, tannins can actually be good for some animals; for example, they can help control microbial growth in the first stomach of cows.

our own alkaloids in the lab, they all came from plants; about 18 percent of plants make them.

Some plant toxins are marvelously specific and wonderfully named. Ricin is the most famous example of a group called "ribosome inactivating proteins," or "RIP" for short. Enough ricin will make you RIP . . . permanently. Remember the ribosome from high school biology? It's the molecular machine responsible for assembling proteins based on the sequence of your DNA. Ribosomes are pretty damn big, by cellular standards: they're made up of seventy-nine proteins and a few thousand-unit-long chains of nucleic acids (RNA). Ricin removes a *single* nucleic acid from the ribosome, which completely and irreversibly inactivates *the entire thing*. And then it keeps on going, clipping other ribosomes' wings, inactivating more than a thousand per minute. Eventually it inactivates enough ribosomes that the cell dies. Let's pause here, because that is ri*donk*ulous: one single molecule of ricin is enough to kill an entire cell. To give you some perspective, one molecule of ricin weighs about 0.000000000000000005 grams; a cell weighs about 400 million times that. Killing an entire cell with one molecule of ricin would be like killing a person with the right leg of an ant.

Unfortunately, ricin is fairly easy to get your hands on, because it's made in appreciable quantities by the castor bean plant. This makes it a popular choice for amateur assassins trying to kill famous people by mail.* Because ricin is so potent and easily accessible, the U.S. Chemical Warfare Service looked into using it as a biological weapon in the mid-1940s. But, luckily for humanity, it's quite hard to turn it into a fine powder, which is what you'd need to kill large numbers of people.

Other plant toxins are, like tannins, much slower killers. Nardoo,

....

* If you've ever seen a bottle of castor oil at a pharmacy, rest assured that unless something in the production process went seriously wrong, it does not contain ricin. (The oil is extracted, leaving the poison behind in the mashed-up beans.)

a species of Australian fern, produces high levels of an enzyme called thiaminase, which specifically breaks down thiamine, a.k.a. vitamin B_1. If you go without B_1 for long periods of time, you end up with a disease called beriberi. Eventually, beriberi will kill you, but not before making you wish you were dead. This is exactly what happened to two British explorers wandering around Australia in 1861: not knowing any better, they made flour from nardoo incorrectly, wound up with beriberi (among other ailments), and slowly died.

Some forms of plant defense are so familiar that we forget their original purpose. You know that warm and cozy pine smell? That's a defense system at work. When insects chew up conifers, the trees respond by oozing out a resin dissolved in turpentine at the site of the injury. The turpentine evaporates (carrying some delightfully smelly molecules to your nose), leaving the hardened resin behind, sealing up the wound and forming what we know as "amber." Often this traps insects. Imagine sitting down for a nice bite of pine tree and finding yourself suddenly enveloped in a sticky golden prison that could become your home . . . for the next 50 million years. Well played, pine tree . . . well played. Some other plants store resin under so much pressure that it can shoot out almost five feet—as if being squirted out of a hypodermic needle—when an insect chews through a vein. Biologists call this the "squirt gun defense."

Latex—yes, the stuff in that beige glove your doctor theatrically snaps on before a rectal exam—is more than just sex rubber. Two researchers who study latex dubbed it "toxic white glue" in 2009, with good reason: depending on the species of plant, it can contain hundreds of different toxins. Latex also contains lots of tiny rubber particles suspended in the liquid, and, like pine resin, that rubber can trap whole insects. But it can also just gum up their mouthparts. Imagine a thousand tiny rubber bands holding your mouth closed; it's like that.

Plants are brutal.

Do plants feel bad about causing all this havoc? There's only one way to find out: we can ask them. Researchers at MIT have recently crossbred the European sneezewort with a MacBook Pro, thus allowing them to access the plant's consciousness. After millennia of human existence, we can finally—

Just messin' with ya.

Plants are incredible beings, but so far, humanity has been unable to get them to literally speak their secrets. So we can't ask a castor bean whether it evolved ricin specifically to kill mammals or to do some important function within the cell, and its toxicity is only a happy accident. But most scientists agree that most plant toxins are toxic on purpose, evolved to discourage insects and animals from eating them. And because all forms of life—and especially insects and mammals—use basically the same molecules to stay alive, almost any chemical or biological weapon that plants make would be pretty much guaranteed to affect more than one species . . . and that often includes us. Frankly, I'm pretty impressed at all the chemical ways that plants can cause every medical symptom under the sun, including (but certainly not limited to) scratching, burning, and redness in your throat and airways, dizziness, vomiting, diarrhea, difficulty breathing, heart failure, coma, and death.

Plants' chemical arsenal can seem overwhelming, unstoppable, even shocking-and-awing, but the animal kingdom doesn't just happily eat the poison and hope for the best. "The plant might be able to evolve a toxin, but then some insects might evolve to overcome that toxin," says botanist Fabian Michelangeli. "It becomes an arms race."

For example, your body has a cyanide detox system based on an enzyme called rhodanese. Many living things have this rhodanese system, presumably to avoid dying if they accidentally eat some plant cyanide.* But it doesn't stop there. Insects and animals can do much more than just try to chemically destroy the toxin. To overcome tannins—which bind to some of the protein in food and prevent it from being digested—many different species, including moose, beavers, mule deer, and black bears, produce proteins in their saliva that sponge up the tannins, preventing them from binding to the food proteins the animal is trying to digest.

Cyanogenic glycosides (remember the "grenades"?) are present in a ton of different plant species, so some insects and animals have evolved extremely creative ways to eat those plants, cyanide and all. Caterpillars of six-spot burnet moths change the way they eat, taking huge bites to avoid mashing up too many plant cells, which would release cyanide. They also have a highly basic (as in "opposite of acidic," not "pumpkin spice latte") midgut, which reduces the number of grenades the Phillips can de-pin per second. And they also eat *super* fast (almost four square centimeters of leaf per hour), which means they also poop super fast, and that limits the amount of cyanide that can be released inside their body.

Caterpillars of a few species of butterfly and moth have also figured out how to safely handle cyanogenic glycoside grenades. But instead of just crapping the grenades out, they store them to use against *their* predators. In one experiment, scientists raised a group of caterpillars on cyanide-producing plants and another group on non-cyanide-producing plants. Then they offered all

....

* If we have a detox system, why is cyanide still so poisonous? Rhodanese needs to get the sulfur from somewhere, and it gets it from proteins. It takes time and energy for us to break down the proteins, create this sulfur, give it to rhodanese, and let it do its work. Enough cyanide can overwhelm the rhodanese system.

the caterpillars to lizards, a natural predator. Lizards ate less than half the number of caterpillars storing cyanide as those not storing it. Sometimes the lizards would take one bite and quickly decide to drop this whole eat-a-cyanide-filled-caterpillar thing. Those lizards shook their heads, opened their mouths wide, wiped their jaw against the floor or their leg, or kept scraping their tongue against their upper jaw. In other words, they behaved as if someone handed them what they thought was a chocolate chip cookie and it turned out to be oatmeal raisin.

Some species of caterpillars, when disturbed, regurgitate a tiny droplet of cyanide-laced digestive juices, essentially saying to potential predators, *There's more where this came from, so think before you eat . . .* The tobacco hornworm, which feeds on tobacco plants, takes up nicotine from the plant and, when a wolf spider attacks it, releases the nicotine as a gas, at which point the spider nopes the hell outta there. (There's an amazing video of this; I've never seen a spider give up on a meal so quickly.*) The scientists who discovered this behavior named it "toxic halitosis," which I think actually undersells it. Instead of expelling bad breath from its mouth like a human does, the tobacco hornworm expels nicotine through about a dozen small openings distributed across its body, essentially enveloping itself in a cloud of toxic gas that is highly unpleasant for the poor wolf spiders.

To fight back against latex-producing plants, some insects will cut a leaf vein, let the latex ooze out, and then turn around and start eating a part of the leaf past the snipped vein. Because that particular vein has been bled dry, no latex comes out where the insect is feeding. Absolutely dastardly.

This arms race between plants and everything trying to eat plants has been going on for hundreds of millions of years. And

....

* Nicotine is more toxic than you think. We'll come back to this in chapter four.

LATEX VEINS

INSECT BITES, AND THE LATEX POOLS

AND NOW THIS PART IS SAFE TO EAT

then there's us. Humans have been living right in the middle of this war for our entire existence, but we've somehow managed to figure out how to eat plants to our heart's content in spite of all the wonderful and creative chemicals plants make to defend themselves. Sure, some of our capabilities are biochemical— rhodanese, for example—but I would argue that most of it is thanks to our ingenuity.

High up in the Andes, 12,000 feet above sea level, a relatively flat, broad plateau called the Altiplano stretches 600 miles long and 80 miles wide from southern Peru almost to Argentina. It's generally cold and dry, with an unforgiving sun. The atmosphere is thinner here, like butter scraped over too much bread. Life is harder. But there are people who have been making do among these hills for thousands of years. Their primary food—sometimes their only food—is the wild potato. You might think of potatoes as "the bag of starch that goes next to my steak." You're not wrong; they *are* mostly carbs. But they also have decent quantities of vitamins, iron, magnesium, phosphorus, and 2 to 4 percent protein. If you happen to be eking out a tough existence thousands of feet above sea level, wild potatoes can be a lifeline. Except for one small problem: most of them are very toxic. They contain all kinds of poisons[*] that, if you eat enough of them, will give you "severe gastrointestinal disturbances," which is medical code for "stomach pain, cramps, toilet-hug-inducing vomiting, shitting your brains out, or any combination of the above." Cooking the potatoes reduces their toxicity, but some of the poisons are not destroyed by heat; even cooked potatoes are not safe to eat. If you're literally starving to death, it's better to keep starving and hope for a miracle than to eat a toxic wild potato.

It's possible that, one day millions of years from now, humans living in the Altiplano will evolve super-effective biological defenses against potato poisons. But that doesn't help you *today*. Luckily, there *is* something you can do right this very moment that would, almost magically, enable you to eat wild potatoes with no ill effects. It's simple, easy, and free. You could do it in your own kitchen. In fact, you could do it outside. It's an old-timey insult: eat dirt. Not just any dirt, but *clay*. And not just any clay. Indigenous peoples of the Altiplano—called the Aymara people— dig six to ten feet below the surface to find *p'asa*, *p'asalla*, or *ch'aqo*, three specific clays that each have their own look, feel, and taste.[†] All three clays work exactly the same way: they act like a sponge, soaking up enough of the potato poisons to make them safe to eat. No matter how you eat the clay—whether you cook the potatoes in a sauce made with clay, or cook them separately and then dip them into a clay slurry (like fries into ketchup)—clays are almost

....

[*] Glycoalkaloids, phytohemagglutinins, protease inhibitors, sesquiterpene phytoalexins, to name a few.
[†] You don't have to actually go dig for these yourself; now, thanks to the magic of capitalism, all three are pretty widely available in Altiplano markets.

magically effective at detoxifying potatoes. It takes only 60 milli-grams of *p'asa*, the most effective of the three clays, to soak up 30 milligrams of tomatine, a glycoalkaloid poison found in wild pota-toes. Ten or fifteen potatoes could be detoxified with a couple tea-spoons of *p'asa*.[*] (The Aymara use much more than this: they seem to feel it's better to eat too much clay than to throw up their meal. Seems like sound logic to me.)

Eating clay (or other minerals) to detoxify potentially dangerous food is arguably the very first thing we did as a species that could be considered "processing": we took something from nature and—before eating or using it—we changed it in some way. At its heart, processing is simply changing nature to suit our needs. Now, if you're thinking, *Eating clay with potatoes isn't processing potatoes; it's just eating two things at the same time,* I understand. Dipping potatoes into a clay slurry might fall just outside the broadest possible definition of processing; it's basically like dip-ping fries in ketchup, if the fries were toxic and the ketchup was the antidote. So let's take another example, also involving the Aymara and poisonous potatoes.

As a kid, I would make annual summer pilgrimages to the Na-tional Air and Space Museum in Washington, D.C. The highlight of these trips was always buying space ice cream, a small rectan-gular lump of freeze-dried ice cream: regular old Earth ice cream that had been simultaneously frozen and dried, leaving the flavor and (most of) the texture, but removing all the water. Freeze-drying something is a pain in the ass. With modern technology, the process goes something like this:

....

[*] Before you rush to buy *p'asa* powder online, let me say: you don't need it. Potatoes you buy in the store are domesticated. The poison has been bred out.

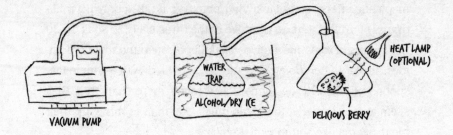

VACUUM PUMP

WATER TRAP

ALCOHOL/DRY ICE

HEAT LAMP (OPTIONAL)

DELICIOUS BERRY

1. Get a strong vacuum pump, some alcohol and dry ice, and some leakproof piping and flasks.

2. Freeze the thing you want to freeze-dry, then put it in a flask.

3. Connect the flask to some pipe, then connect the other end to a second flask.

4. Dunk the second flask into an alcohol/dry-ice bath.

5. Connect the second flask to a vacuum pump.

6. Turn on the pump, and let run for at least twelve hours.

7. A few hours in, heat the flask gently with one of those red lightbulbs that keeps you warm after a shower.

8. Wait a few more hours, and finally . . .

9. Enjoy your NASA ice cream.

Here's how this works: the vacuum pump lowers the pressure to almost zero, which causes the frozen water in the ice cream to start to evaporate—without melting. Heat from the light helps this process along. As the water vapor enters the second flask, it freezes. Eventually you end up with freezing cold, bone-dry food. Essentially what you're doing is using low pressure, extreme cold, and gentle heat to remove solid water (ice) out of a frozen food without melting the food first.

Freeze-drying food seems like a modern technology. But the Aymara figured out how to freeze-dry potatoes without pumps or pipes or a freezer. Here's how they do it:

1. **Get some wild toxic potatoes.**

2. **Freeze the potatoes by leaving them outside overnight at high altitude.**

3. **Trample the frozen-solid potatoes like a French winemaker tramples grapes.**

4. **Put the trampled potatoes in a loose wicker basket, put the basket in a stream or creek, and leave for a few weeks.**

5. **Put the potatoes on your front doorstep and let them freeze overnight and dry out in the daytime, squeezing occasionally, then leave for another few weeks.**

6. *¡Y voilà!* **Freeze-dried potatoes.**

This method is astonishingly similar to modern-day techniques. Instead of a vacuum pump, the Aymara use their environment: at high altitudes, the pressure is low. Instead of warming lightbulbs, the Aymara use the sun. The Aymara method is even a bit more sophisticated than modern methods: trampling the potatoes and then leaving them in running water leaches out about 97 percent of the toxins in the wild potatoes.[*] And not only is the

....

[*] It also leaches out almost all the protein and many of the vitamins and minerals. But hey, life is full of trade-offs.

final product edible without gastrointestinal distress, it's much more storable. Fresh potatoes might last for a year; leached and freeze-dried potatoes can last for *twenty* (some people say indefinitely). If you were a member of a society like the Aymara, having a ready supply of edible stored carbs that would last you through a two- or three-year famine might just be the key to your survival.

The historical record is not clear as to whether or not this is the world's first processed food, but it *is* clear that this is processing: taking something from nature and changing it to suit our purposes—in this case making it nontoxic.

Let's look at a much more common crop: cassava. You might know it as manioc or yuca. Depending on where you live, cassava is either an ingredient on a fancy menu or an essential part of your everyday calorie count. Says Ros Gleadow, an Aussie plant scientist: "Cassava is very, very important for feeding people. We don't eat it much in Australia, but a billion people around the world eat it every day." Cassava plants are a farmer's dream: they're easy to propagate, can tolerate crappy or untilled soil, and are extremely low-maintenance. They're drought-resistant, and you can leave the roots in the ground for up to three years after maturity, which basically gives you plant-based insurance against famine. If you were into cheap puns, you might call it an insurance plan . . . t.

There is, of course, a catch. If you haven't guessed already, here's a hint: How do you think the plant manages to keep its highly nutritious, starchy tubers uneaten by passing animals or insects for three years after they reach maturity? Yep: cyanide. As we've already seen, lots of plants produce cyanogenic glycosides—in fact, two-thirds of all crop plants have at least one part that produces cyanide—but some varieties of cassava that are eaten throughout the world produce enough cyanogenic glycosides in their tubers to kill an adult human.* And, unfortunately, simple techniques like roasting or boiling the cassava do nothing to get rid of the cyanogenic glycosides. But if you process the cassava just right, you *can* get the cyanide out.

Let's revisit our grenade metaphor: Remember that plants store the grenades (cyanogenic glycosides) separately from the Phillips (pin pullers), and that when some insect comes by and starts wantonly mashing up plant cells, this brings together the Phillip and the grenades. Phillip pulls the pins, the grenades go off, and cyanide is released.

Paradoxically, the first step in any cassava detoxification method is to pull the pins on *ALL THE GRENADES!*, generating as much cyanide as possible—but outside of a human body. For example, you could physically rip apart the plant cells by grating the living crap out of the tubers. Or you could let bacteria and fungi chew up the plant cells for you via fermentation. Once the cyanide is generated, the second step is to remove it. Luckily, it dissolves in water or evaporates fairly easily. So once you have your cassava mash, you can strain it, boil the water away, or lay it out in shallow wide pans and leave it in the hot sun for a few hours. In South America, cassava flour or mash is often strained using a special woven device called a tipiti: imagine a Chinese finger trap, but four feet long. You load it up with cassava mash, hang one end from a rafter or tree branch, then use your body weight to pull the other side down. Cyanide-laced water is pressed out, leaving safe-to-eat cassava. Pretty damn ingenious.

Transforming toxic plants into edible food is the bare minimum amount of processing that we, as a species, can do. We've been doing this for much of our history, and many people still rely on

....

* One unfortunate case involved two people who died after eating cyanide-containing cassava . . . served at a funeral.

this basic form of processing to this day. We eat potatoes with clay or leach and freeze-dry them; we bake acorn bread with clay; and of course the holy grail of food processing is changing the very genome of the plant by selective breeding so that they aren't toxic in the first place.*

You might be asking: Why don't we just leave toxic plants alone and eat nontoxic stuff instead? That's perfectly reasonable when there are lots of nontoxic foods around to eat. But if your source of perfectly safe, nontoxic sugar/fat/protein dries up, you better have a backup plan . . . or you'll starve to death. It's pretty simple, brutal logic: the more things—toxic or not—you can turn into a meal, the more likely you are to survive.

But immediate survival isn't the only reason we process stuff.

....
* But that's another book.

CHAPTER THREE

MICROBES ARE TRYING TO EAT YOUR FOOD

This chapter is about two dead cows, honey, water, the bacteria living on your shower curtain, Martha Stewart, small green insects, the Owens Valley Paiute, sugar, honey, and blood.

We're going to do a thought experiment, and it's going to get weird.

Imagine two dead cows lying right in front of you.

The dead cow on the left (Berta) you'd like to make disappear as quickly as possible: she's evidence, and you need to get rid of her stat. The dead cow on the right (Wilhelmina), you'd like to preserve as long as bovinely possible. We're not talking days or months here. You'd like to preserve Wilhelmina until humanity's long-predicted dystopian future comes to pass, when we're all eating Soylent Green.

Chemically speaking, Berta is the easy nut to crack. According to the movie *Snatch*, chopping her up into bite-sized pieces and feeding her to pigs is the surest way to get rid of every trace, but honestly you could leave her body almost anywhere in the world and she'd decompose pretty fast. Preserving Wilhelmina would be a lot harder than getting rid of Berta. Depending where you are on the planet, you could get lucky. For example, if you happen to live near the North Pole, you could just leave her outside, in what is essentially a giant natural freezer. Eventually, of course, Wilma's body would also decompose, but it would take much, much, muuuuuch longer.

Why are we disposing of dead cows in our head? A couple reasons: first, it's less weird than disposing of dead humans; but more important, because most food starts with (or includes) murder. Almost everything you have ever eaten or will ever eat was once a living, respiring thing, or part of a living thing. Proteins, fats, carbohydrates, and fiber—which make up the majority of your diet and without which you would starve to death—don't just spring from the earth fully formed. Plants make them; animals eat those plants. Then *we* eat those plants and animals. This isn't to make you feel bad (or . . . good?). It's to reacquaint ourselves with the idea that when we eat, we're eating dead bodies of plants and animals. This is important, because lots of other things are trying to eat dead bodies, too.

At one time during human history, back when we were basically eating food within a few hours of killing it, we were competing against hyenas, vultures, flies, and other creatures visible to our naked eyes. Once we got the crazy idea into our heads that we wanted to kill something *now* and not eat it until *days or weeks later*, we started a race between ourselves and a zoo of invisible microbes,[*] each trying to outrun the other to see who could eat a piece of bread or fruit or a chunk of Berta first. Let me be clear: the microbes always win in the end. "They were here before us,

....

[*] Calling it a "zoo" doesn't do justice to the number of microbes involved. If you were to take every single zoo in the world and add up all the animals in them, and then add approximately a bazillion to that number, you would have the number of microbes on a small patch of rotting meat in your fridge.

and they'll be here after all humans are gone. They will prevail," says food scientist Susanne Knøchel very cheerily. Why? "They're everywhere. Even in places where fifty years ago people had no idea there were microbes, we find them." They're floating around in the atmosphere, hitching a ride on the dust in your house, stuck to your showerhead, camping out in your shower curtains (you know exactly where I'm talking about), and colonizing pretty much your entire kitchen. And, of course, they're inside of *you* (and Berta), fermenting away a portion of the "indigestible" plant fiber you eat. In fact, you have about as many bacterial cells in your intestine as you do human cells in your entire body. Our gut microbes (part of our microbiome, the collection of microbes that live on and in us) are important for our survival in ways that we're still figuring out. But our relationship with these creatures is only a temporary truce: while we're alive, we give them a warm, wet habitat and lots of food, and they give us energy and help protect us from their harmful cousins—but the moment we die, these little creatures will turn on us and devour us from the inside out.

And it's not just your microbiome: depending on how and where you die, a wide variety of microbes and other living things will happily eat the proteins, fats, carbs, vitamins, minerals, and every other element that made you who you were and use them for their own benefit. Eventually you will disappear. Your body's decomposition is an all-*they*-can-eat buffet. Don't feel bad: this is true of almost every living thing. Once something dies, it usually becomes food for something else. Berta is no exception. Her viscera and soft tissue would be devoured first. Her skeleton would last longer, but there are life-forms that eat bones, too. Over time, she would be completely broken down and, having fed an almost uncountable number of bacteria, fungi, molds, insects, animals, and plants, her atoms would be spread throughout billions of other living creatures on the planet.

This is called decomposition, also known as rotting, and it's the completely natural, totally normal thing that happens to living things' bodies after they die. That's why it's so much easier to get rid of Berta than it is to preserve Wilhelmina.

Doesn't mean humanity hasn't tried, though.

Suppose you wanted to preserve Wilhelmina for as long as possible. To do that, you would have to prevent microbes from devouring her *and* you would have to stop the life processes within her own cells. The best way to do that is to embalm her. One of the most impressive embalmers is also one of the simplest molecules: formaldehyde. It has just one carbon, one oxygen, and two hydrogens:

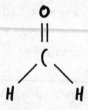

But don't let the simplicity of the molecule fake you out. Formaldehyde is extremely (chemically) vicious and promiscuous. See the carbon? It's what chemists call electron deficient, because the oxygen is pulling electron density away from it.

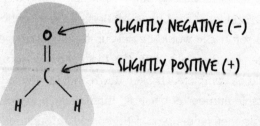

So the carbon ends up with a little bit of a positive charge, which means that it's attracted to parts of other molecules that have a little bit of a negative charge.

Where might there be such molecules? You are made of them. Many molecules in every single one of your cells qualify: the proteins that fight infection or help store and copy your DNA, some of the fats that make up the barrier between your cells and everything else in the universe, some of the carbs you burn for energy or store for later, and even the RNA and DNA that make up your genetic code—almost all of these molecules have some areas within them that are a little more negatively charged and are thus able to react with formaldehyde. If and when formaldehyde collides with these slightly negatively charged areas in other molecules, the two molecules—formaldehyde and, for example, a protein—can join to become one. And the reaction doesn't stop there. Formaldehyde, once bound to a protein, can react a second time in essentially the same way, binding to any other molecule that has a little bit of concentrated negative charge. This is often another protein or a strand of DNA.

So, at the beginning of this whole process, you might have three molecules: a huge protein, a super long strand of DNA, and a teensy-tiny molecule of formaldehyde. At the end, those three molecules become one, linked together by the Little Formaldehyde That Could.

Embalming with formaldehyde is this reaction but on a *massive* scale. Imagine spraying 6 million gallons of superglue onto New York City at rush hour.[*] Within minutes, people would be stuck to

....

[*] I actually did this calculation. It goes something like this: it takes about 4.5 grams of formaldehyde to completely "fix" 100 grams of the average water-soluble protein. Using this same ratio, it would take 6.75 pounds of superglue to completely immobilize a 150-pound person. New York City has a population of about 8 million people, so you'd need about 54 million pounds of superglue to fix them all. Cyanoacrylate, the main component of most superglues, has a density of roughly 1.1 grams per milliliter, which means 54 million pounds would be a shade under 6 million gallons.

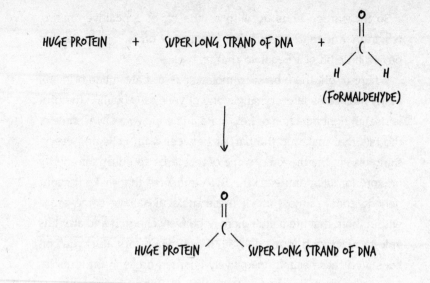

HUGE PROTEIN   +   SUPER LONG STRAND OF DNA   +   (FORMALDEHYDE)

HUGE PROTEIN   SUPER LONG STRAND OF DNA

sidewalks, streetlamps, signs, hot dog vendors, and one another; cars, buses, trucks, and trains would be stuck to the streets and rails; and everyone on a United flight would feel . . . just as stuck as before. All the people, cars, trucks, buses, trains, etc., would still be vibrating in place, struggling against the formaldehyde glue, but all their normal, long-range motions would stop in their tracks.

Life needs motion. Molecules have places to go and stuff to do; stopping that motion brings cellular life to a grinding halt. From the perspective of a bacterium looking for a meal, formaldehyde turns an orgy-inducing amount of food into a gigantic, useless museum. It is the ultimate preservative.

You might guess that a chemical that stops life processes would also be toxic. You'd be right, though formaldehyde is not as potent as cyanide. About 12 to 20 grams would be necessary to kill a human adult, and death by formaldehyde would not be pleasant.

Before it was used as an embalming fluid, formaldehyde was used as a tanner—in other words, a liquid that turns animal skin into leather. In one case of suicidal formaldehyde poisoning, the doctors noted that the patient had died of massive lung damage and a "leatherlike thickening of the gastric wall." Given its toxicity, there are a shocking number of cases in which live humans were accidentally dosed with formaldehyde, in a wide variety of orifices, usually because someone was being a doofus. Formaldehyde has been accidentally injected into the eyelids of a three-year-old child and a fifty-nine-year-old woman, and into the gums of a twenty-three-year-old man (some genius dentist let an unsupervised undergraduate perform a tooth extraction); it has been accidentally infused into the veins of dialysis patients, which must feel like you're being burned at the stake; and, incredibly, a patient survived after receiving a 100-milliliter *enema* of 4 percent formaldehyde. In perhaps my favorite case of accidental formaldehyde poisoning, a surgeon accidentally injected formaldehyde directly into a patient's kneecap. The source of the formaldehyde was a small vial in which a piece of that very same patient's knee had been preserved; it was intended as a gift, presumably to commemorate a successful surgery. So much for that.

Anyway, if you were to preserve Wilhelmina in a gigantic airtight vat full of formaldehyde, she would last . . . well, no one actually knows how long she would last. Formaldehyde was first used to embalm a cadaver in 1899, and that cadaver is presumably still (not) going strong, so Wilhelmina would last at least 120 years, and—given what we understand about the nature of formaldehyde fixation—probably a lot longer.

So we can now set up the theoretical limits of our dead cow thought experiment; what I'll call the Berta-Wilhelmina continuum:

BERTA		WILHELMINA
WARM, WET ENVIRONMENT		FORMALDEHYDE EMBALMING
LIFE LIVED VICARIOUSLY		LIFE STOPPED IN ITS TRACKS
FAST ROTTING		ETERNAL PRESERVATION

Food goes bad because of *life*: the life that lingers in its cells after the organism dies and the life that takes over the body of the dead. Preventing that life prevents decomposition.

Because this continuum applies to all dead stuff, not just cows, and because all food was once alive, we can add one more line:

FOOD ROTS		FOOD LASTS FOREVER

Everything on the left is the same: Berta's fast decomposition is the result of lots of microscopic life-forms doing their damnedest to gorge themselves and reproduce as much as possible.[*] Berta could have been human food, but the microbes got to her first, and she was spoiled. Likewise, everything on the right is the same: Wilhelmina was preserved because formaldehyde comes as close as chemically possible to stopping the motion of life within each and every one of her cells, as well as within the cells of all the organisms that were chowing down on her.

Before preserving food was a science, it was an art: finding a happy middle ground between Berta and Wilhelmina, allowing

....

[*] Not all food spoilage is caused by microbes. Some is a result of chemical reactions that happen in the food itself, without any help from other living things. For example, unsaturated fats like olive oil can go rancid all by themselves, thanks to the double bonds in the fats reacting with oxygen in the air.

just enough life—or the right kind of life—to keep your food edible, but not so much life that the food decomposed. Preserving a food necessarily involves changing it—just enough to stop or slow the life within its cells, or make it uninhabitable to uninvited microbes, but not so much that it becomes a museum. To see some of the weird and wonderful ways people have devised to preserve food, you need only travel to that whopping warehouse consecrated to the free-ish trade of food, the swindler of cilantro and the purveyor of peas, the messenger of milk and the ode to oatmeal: the supermarket. Some of the preservation techniques are simple: fresh fruits and vegetables are chilled to slow down molecular motion and thus spoilage. The frozen section is a more extreme version of this. Some techniques are complicated and largely invisible; for example, a technique called "manothermosonication"* can be used to preserve milk and orange juice. But most of the techniques used to preserve foods in the modern grocery store are very old, of unknown provenance, and incredibly—almost mysteriously—effective. Chief among these is simply drying food out.

People have been drying food for millennia, possibly even before they learned to cook it. The grocery store is full of preserved foods that are very obviously dry—flour, cocoa, powdered milk, potato/tortilla/veggie chips, oats, nuts, etc.—but it's actually also full of preserved foods that seem wet but aren't. Jam, molasses, corn syrup, condensed milk, butter, and honey are actually much drier than they look.

Drying is the removal of water; so let's talk H_2O. If you haven't thought much about water, it might seem . . . well, banal.

....

* Mano-thermo-sonication is the simultaneous application of high pressure, high temperature, and loud and destructive sound waves, like when an obnoxious bro asks you out in the middle of hot yoga.

Quotidian. A jaded chemist might even say boring. Unlike most chemicals that typically capture the Internet's imagination, water is clear, colorless, tasteless, odorless, and just about every other "-less" you can think of. You can leave it for years and it won't change. It's almost nontoxic—unless, of course, you drown in it—and it's not (too) corrosive. And yet—despite these Internet-unfriendly characteristics—water is essential to every single form of life we have ever discovered.

For this next part, it's helpful to clear your head of whatever images pop into it when you read the word "water": streams, rivers, glaciers, pee, oceans, rain—all of that imagery will not only not help your understanding but will actively hinder it, because it makes you think of water as *one fluid.* You've seen this illustration before:

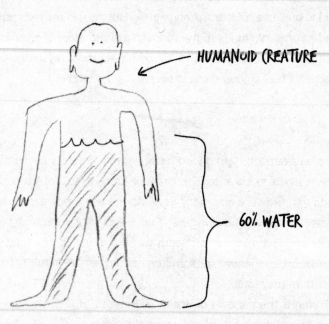

HUMANOID CREATURE

60% WATER

It's meant to show how much water is in your body. Unfortunately, the implicit message is that water fills you up like it does a cup. But when you look at water's behavior inside of living things, at the scale of individual proteins or DNA, nothing could be further from the truth.

Imagine a freewheeling collection of tiny elbow-shaped robots, each embedded with two tiny magnets that attract or repel the magnets within other robots, and each robot having the ability to cut off one-third of another robot and attach that piece to itself, or lose one-third of itself to a peer quite easily; these two abilities allow the robots—which, in a thimble-sized volume, outnumber stars in the known universe—to form, unform, and re-form extended three-dimensional networks made from trillions of trillions of robots, billions of times per second.

ZOMG, right?

When you stop thinking of water as a clear liquid of nothingness and start thinking of it as a (mostly) benevolent, extremely hyperactive but non-sentient mechanical civilization, it's easier to understand why water is so crucial to the cellular machinery that keeps us—and the bacteria intent on spoiling our food—humming.

Water's magnet-like behavior is not unique.[*] In fact, most molecules behave as though they were embedded with tiny magnets, usually more than one per molecule. Chemists call these molecules polar. DNA is one such molecule. Polar molecules can interact much more strongly with each other than they can with nonpolar molecules (molecules without strong discernible magnet-like behavior). If you missed all of that, don't worry; the point is that water and DNA attract each other, so much so that DNA is actually *coated* in several layers of water molecules.

Now, here's where you have to shift your mind-set. Think of a human body coated with water: the coating is a smooth, shiny,

....

[*] Confusingly, despite individual molecules behaving as though they were magnetic, a glass of water is *not* magnetic. (Try waving a magnet around near water: nothing will happen.) Water molecules behave like little magnets because of *electric* fields created by the nuclei and electrons of those molecules. Weird, I know. Blame physics.

slippery surface. Not so at the molecular scale. Imagine billions of tiny robots loosely attaching themselves in a single layer to a strand of DNA. Now imagine a second layer of tiny robots attaching themselves to the first layer. Now imagine a third. That's how water molecules hydrate DNA: as a vibrant colony of bees might swarm its queen around a beekeeper's neck—constantly detaching and reattaching, moving, but retaining the underlying shape of the keeper; creating out of many, one.

Like the human beneath the bees, the underlying structure of the DNA shows through, one or two layers of water molecules away. This means that proteins that need to read DNA—for example, to copy it when a cell divides, or to fix damage—can sense the underlying sequence of the DNA *without having to fully attach themselves to it*. And because the water molecules are able to form and break attachments relatively easily, DNA-reading proteins can scoot along the DNA-shaped water layer that surrounds the actual DNA without wasting energy by stopping and binding to the DNA itself.

This is just one example of all the amazing things water can do. Water is definitely *not* sentient, but sometimes it seems like it just . . . barely . . . could be . . . No other molecule—that I know of, and I'm friends with many—can do all the things water can. As biophysical chemist Bertil Halle said in 2004: "There is only one way to make proteins, there is only one way to do photosynthesis, there is only one way to store and transfer information. All forms of life use the same molecular mechanisms." That means all forms of life (that we've found so far) need at least some water to live. And that's why removing it preserves food.

So it's not surprising that lots of things in the High Temple of Food Preservation, a.k.a. the grocery store, are dry. Almost everything in the chip aisle is preserved by drying. (In this case, the method is submersion of the food into a complex mixture of liquid fats heated to about 302°F (150°C), at which temperature most water within potato or corn cells boils off—also known as "frying.") Most cereals and snacks with weird shapes (think cheese puffs) are also heated (and thus dried) to the point of being crispy. Even foods in the frozen section are dryish, at least as far as microbes are concerned . . . Freezing is a one-two punch: not only does it slow down all molecular motion—and therefore all life—but frozen water takes on a highly rigid, crystalline structure that is capable of bursting cells and incapable of supporting microbial growth.

All life we've found so far needs water, but different kinds of life need very different amounts of water. If you wanted to make sure *not a single living thing of any species* is able to grow in your food, you would dry it out completely, wringing every last water molecule from its cold, dead body. Unfortunately, it turns out that the only way to do that is to burn every last cell to a literal crisp, until all that's left is ash. Fortunately, you don't need to remove every molecule of water from a food to render it inhospitable to the forms of life that would spoil it or make it dangerous to eat. You only need to remove *enough* water. How much is enough? Well, that depends on what you're trying to kill.

Let's say your target is *Escherichia coli*, also known as the "Something pooped in this" bacterium, because wherever you find it, it almost certainly came from a mammal's intestine.[*] Turns out, *E. coli* is not very good at living in even marginally drier environments. That's why whenever you hear about an *E. coli*

....

[*] *Escherichia* comes from Theodor Escherich, the German doctor who discovered it in 1886, and *coli* comes from the Latin *colon*, which means exactly the same as in English. So the name of this bacterium is effectively "German doctor's poop-tube bacterium." Who said scientific names were boring?

O157:H7 outbreak, the culprit was usually something with plenty of water in it, like beef, dairy, or fresh fruit and vegetables. Yeasts are typically hardier than bacteria, and molds are hardier than yeasts, but there is a threshold level at which nothing can grow. The dry spices in your spice cabinet, boxed pasta, cocoa powder, powdered milk, and potato chips are all below this threshold water level. So are a surprising number of seemingly "wet" products, like honey.

Honey is the product of a lot of processing. In fact, it may be the original processed food; just not by humans. Bees spend the summer collecting nectar, which is 30 to 50 percent sugar, and then they concentrate that nectar to roughly 75 percent sugar, add a few molecules of their own, *et voilà*: a magically sweet food that's incredibly energy dense and yet highly inhospitable to microbes, which keeps it edible through the winter.[*] That inhospitableness is thanks in part to its dryness.

Objectively, this makes no sense. Most honey is a pourable liquid. How is something that looks like slow-motion water *dry*? Well, "dry" doesn't just mean how much water is in a food; it also means how much water can that food *give* to a microbe that wants to live in it. Honey is about 15 percent water (which puts it in the same realm of water content as things like rice or marzipan), 10 percent other stuff, and about 75 percent sugars, mostly fructose, glucose, and maltose. Let's look at the chemical structures of these sugars:

See all those "—OH" groups attached to the sugars? Glucose and fructose each have 5; maltose has 8. You can think of each one of those as behaving like two tiny little magnets, each one able to attract a water molecule. And just like water hydrates DNA

....
* Let's be honest, we all have that jar of honey from 1997 that's still perfectly fine.

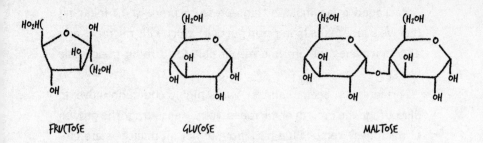

FRUCTOSE   GLUCOSE   MALTOSE

in multiple layers, it hydrates sugar in multiple layers. Martin Chaplin, one of many scientists who has devoted his entire life to understanding water, found that one molecule of glucose was able to attract and mostly hold on to twenty-one molecules of water. And the key here is: the more strongly all those sugars in honey hold on to water, the less likely that microbes are to pry those water molecules away, so the less able they are to grow.* So even though honey is a free-flowing liquid and is made of about 15 percent water, not nearly enough of that water is available for microbes to use, live, and reproduce. Jams, jellies, and preserves operate on the same principle, using sugar's ability to bind water, thus preventing microbes from using it.

Other preservation techniques are somewhat more adventurous.

....

* If any new moms are reading this, you might be wondering, *Wait a hot second. If no microbes can live in honey, how come I shouldn't give honey to my baby?* Answer: Because honey might contain spores of *Clostridium botulinum*. A spore is what a bacterium becomes when conditions are too brutal for it to live normally but not brutal enough to fully kill it. They can't live or reproduce, but they're not dead, either: they're biding their time, waiting for the moment when chance or fate moves them from a hostile environment (for example, honey) to a friendly one (your baby's intestine). Once in friendlier climes, spores un-spore themselves and become live bacteria. In the case of *C. botulinum*, this is decidedly not good, because, as part of its normal cellular functions, *C. botulinum* bacteria happen to excrete literally the most toxic substance known to man, a protein roughly hundreds of thousands of times deadlier than cyanide. But who's counting.

Packaged guacamole is "high-pressure processed," meaning that we squeeze the living crap out of it, which kills microbes and also inactivates the enzyme responsible for turning guacamole black.

Fermentation seems familiar but is highly counterintuitive: to discourage the growth of microbes, you . . . encourage the growth of microbes? Yep. A little background: not all microbes are created equal. There are literally millions of species of microbes on this planet; most of them are completely harmless to us, and some are extremely beneficial. Fermentation is basically encouraging good microbes like lactobacillus to come have an orgy in your food. Lactobacillus eats sugar, excretes lactic acid, and reproduces at speeds that make rabbits look like nuns. In the rich tradition of ancient Rome, they eat, drink, multiply, vomit, and then pass out. This disgusting (or wonderful) bacchanalia turns a perfectly happy microbial home like milk (with a very livable pH of 6.5-ish and *excellent* public schools) into a corrosive hellswamp, 100 times more acidic than milk and absolutely unlivable to the vast majority of all other microbes, including and especially the ones that make us sick. This corrosive hellswamp goes by the common name of "yogurt" and is thankfully unlivable to, for example, *C. botulinum*. Lactobacillus is just one example of many; fermentation is responsible for not just yogurt but also cheese, sour cream, beer, wine, vinegar, sauerkraut, kimchee, bread, and countless other foods.

And then, of course, there are canned foods, the staple of doomsday preppers everywhere. Canning foods denies microbes their access to oxygen, which you would think should be enough to kill them all, but you would be wrong. Our old friend *Clostridium botulinum* actually thrives in oxygen-free environments, which means that once foods with a pH higher than 4.6 are canned, they have to be heated to a specific temperature and held there long enough so that the probability of finding *C. botulinum* in a can is one in a billion.

———

In her 2015 book, *The Food Babe Way: Break Free from the Hidden Toxins in Your Food and Lose Weight, Look Years Younger, and Get Healthy in Just 21 Days*, Vani Hari wrote this:

> As you stroll down the aisles of the grocery store, start thinking about the shelves of boxed, canned, jarred, and packaged foods as caskets holding dead food. It's all embalmed with preservatives that will make you feel dead, too.

Comparing food preservation with corpse embalming makes for A+ shock value! "Embalming" makes you think of a funeral home, dead people, and *Six Feet Under*, none of which you want to associate with a Cheeto, a slice of sausage, or a dollop of mustard. I have to say, the Food Babe is . . . kinda right! In my opinion, preserving food isn't just *like* embalming; it *is* embalming! Well, not full embalming. More like . . . *diet embalming*. Or *just enough embalming*. After all, if you don't embalm, bodies rot—whether plant or human. We've intuitively realized this for a long time, and the result is some pretty cool stuff. For example, if you've ever enjoyed dried and salted cod, you're eating the corpse of a fish preserved in approximately the same way as the ancient Egyptians preserved the corpses of their kings, except, instead of table salt, the ancient Egyptians used natron. If you've had the privilege of eating a particular kind of ham called *jambon de Paris*, sold only in France, you're enjoying the salt-embalmed body of a pig. In this particular type of ham, the salt solution is infused throughout the pig by injecting it into its venous system in exactly the same way as formaldehyde is injected into the venous system of human bodies in funeral homes across America.

And, of course, there's honey. Because honey itself is preserved, what easier way to preserve something else than to drop it in a jug of honey? Everyone got in on this: the Chinese, Indians, Egyptians, Greeks, Romans, and First Nations peoples used honey to preserve everything from seeds to wildflowers to strawberries to dormice. Yup, in the Middle Ages, people would catch these delicate little rodents in their hovels, preserve them in honey, and eat them at their leisure. (Why hunt if the food will come to you?) And when one of the most celebrated military commanders of all time, Alexander the Great, was laid to rest, he was temporarily embalmed in honey for his postmortem journey from modern-day Baghdad to modern-day Greece.

Now, of course, the connection between preservation and embalming is not exact and seamless; it's more like a chemical metaphor. For example, no one has ever, to my knowledge, tried to fry a human body to dry it out, thus preserving it like the sliced and fried body of a potato plant's carbohydrate storage organ. Nor would I say that yogurt is embalmed, although its preservation method relies on the same goal as embalming: preventing microbial growth. Nevertheless, to confirm that I wasn't wildly off the reservation on these types of comparisons, I called a scientist who studies how human and animal bodies decay: Dawnie Steadman. Steadman directs the aptly named Body Farm, a three-acre plot of land where they lay dead bodies out in nature and watch them decompose, to help advance forensic science. When I asked her if there was a huge difference between a decaying human body and a decomposing steak, she said, "No, I think there are obvious similarities. Rotting meat is rotting meat."

And embalming is one specific way to prevent meat—or a plant—from rotting. So in that sense the Food Babe is kinda right. But what's much more important than *whether* something is embalmed is *how* it's embalmed. You'd never want to eat a museum specimen embalmed in formaldehyde. But why that should make you wary of a cucumber "embalmed" in a mixture of water, acetic acid, and salt—also known as a pickle—is beyond me. Whether you're salting a Pharaoh or a fish; or dunking a dormouse or Alexander the Great into honey, it's amazing how similar the chemistry is.

Some preserved foods *are* embalmed corpses—not perfectly embalmed, of course; not to the point of *never* decomposing. Just enough to last you through the winter.

———————

So preservation can yield all kinds of creative foods, which leads me to the next reason we process stuff: for fun. If you're reading this book, you probably associate food with fun: trying a new twist on a recipe, exploring different cuisines, experimenting with weird ingredients. But food and fun have not been partners for very long. Our ancestors had no cooking shows, no intricate recipes, no sous vide, no molecular gastronomy. A prehistoric Bobby Flay wouldn't have had much to do.

But . . . there may have been a few exceptions: cracking open a bone to get at the soft, fatty marrow inside or licking just the right rock for a salty hit. Though there's no way to know for sure, I'd be willing to bet my last Necco wafer that the most delicious, intoxicating, and flat-out fun food of prehistoric times was: honey. If you happened to live thousands of years ago, before humanity figured out how to crystallize sugar from sugarcane (around the first century CE), honey would have almost certainly been the sweetest thing you had ever—or would ever—taste.

Honeybees spend a lot of time and energy building hives, making lots of little baby bees (larvae), feeding them honey, and eating the honey themselves over the winter. If you're a bee, the hive is your home, your source of energy, and the foundation of the next generation. If you're not a bee, the hive is an incredibly tempting food. Honey is chock-full of sugar, and the larvae have almost as much protein per gram as beef, and even more fat. It's no wonder that hive defense is critical—and bees can be pretty creative defenders.

Let's say an ant tries to walk into a hive. Bees will literally blow them off the landing with air currents created by fanning their wings about 275 times per second. Hornets are tougher to defend against. Some species of hornet hunt adult bees, usually when they're returning to the nest laden with honey. Hornets are pretty much un-stingable (they have a tough outer shell), so the bees have to get creative: about fifteen to thirty of them try to grab the hornet and form a ball of living bees around it. If they can manage that, they all shake their butts together, heating themselves—and the trapped hornet—to over 110°F (43°C), killing it. Some species of hornet are heat resistant, so in addition to heating it, the bees block the movement of its abdomen, suffocating it to death, sort of like if a sumo wrestler were to sit on your lungs. When defending against larger creatures, like bears or humans, bees will absolutely sting you.[*] But most of the defenders won't actually sting. They'll harass you: they'll fly right into you, buzz loudly, bite you, and even pull your hair (one at a time, but still!). Basically, they'll try to make your honey-gathering experience as horrible as possible.

Why? Because honey is amazing. It's probably the most calorie-dense food in nature, it's conveniently packaged in a central (though maybe not easily accessible) location, and it comes with a side of bee larvae packed with protein and fat. Also, in case you haven't noticed, it's frickin' delicious.

....

[*] Once a bee stings you, it actually doesn't die immediately. De-stung bees can live for another one to five days.

Now, I'm of the opinion that figuring out crafty ways of getting honey, as early humans did, is *not* processing; it's just stealing another animal's processed food. Stealing honey is all very well and good . . . if you live near a hive. If you don't, you had to find another way to get your sugar high.

Remember that plants are constantly pumping what is essentially syrup from the leaves to the rest of the plant via sieve tubes buried deep within their tissue. If you want access to this sugar rush, you cannot just take a bite of a plant. Leaves, shoots, stems—in other words, the parts of the plant that house most of the sugar superhighway—are not sweet. (Think celery stalks.) That's because when humans take a bite of plant with our gigantic gnashing teeth, we're not just getting the sieve tubes; we're getting all the other parts of the plant that *don't* have a constant stream of sugar whizzing through them, and that tends to cancel out the sappy parts. We're also getting bitter chemicals the plant makes specifically so they don't taste good. Unfortunately, we just don't have the delicate machinery required to dip into a plant's sugar superhighway. But there is a creature that does: the humble little aphid.

Aphids, also known as plant lice, are quite small, usually green,

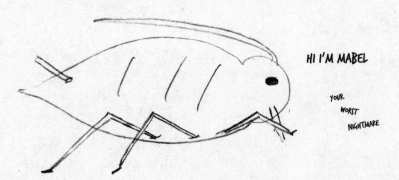

HI I'M MABEL

YOUR
WORST
NIGHTMARE

and absolutely terrible for plants. We'll start our story with a single lady aphid—let's call her Mabel—landing on a plant. Mabel is about 5 millimeters long, but she's big for an aphid. Most species are about 2 to 3 millimeters long. Once Mabel finds a spot she likes, she spits out a small bead of saliva that quickly hardens to the consistency of peanut butter. As it's hardening, Mabel unfurls her "stylet," which is kind of like a hypodermic needle, except it's flexible and has two channels instead of just one.

The stylet is basically Mabel's mouth: her face just sort of stops being a face and starts being a long, flexible needle.

Mabel penetrates the gel saliva that she's just spit out with her hypodermic needle–face, and soon the tip of her stylet arrives at the surface of the plant. Unlike the metal needles that doctors jab you with, Mabel's stylet doesn't punch through plant cells; it worms its way *between* them. Mabel pushes her stylet into the plant in gentle pulses: before each pulse, she spits out a small glob of gel saliva, then penetrates it, and when the tip of her stylet pokes out the other side of the glob, she spits out another glob, penetrates that one until her stylet tip comes out the other side, and so on. These globs of gel saliva harden, creating a sheath that protects (and lubricates) her stylet as she pushes it between plant cells, farther into the plant.

Every so often, Mabel needs to get her bearings. Her stylet doesn't have eyes—it has no way to know where it is inside the plant—so she pokes the tip of her stylet into a nearby cell. Once inside, she takes a "sip" of the cell's contents. In other words, she sucks up some of the cell's guts into one of the two channels in her hypodermic needle–face and "tastes." We don't really know what "tasting" is like for Mabel, but we think she's checking to see how sweet or sour it is. If it's not sweet enough and/or too sour, she retracts her stylet, changes direction, and moves on, deeper into the plant. Eventually she penetrates the holy grail of plant anatomy, the sugar superhighway that is a sieve tube.

As you might guess, plants do not want to be penetrated. Especially not in the sieve tube, because they know what's coming next: large-scale theft of the sugar they've worked so hard to create. Plants are not ungenerous. They have no problem making a fair trade with an insect or an animal, something along the lines of

HEY! YOU, THING THAT CAN MOVE! I'M STUCK HERE, BUT I'VE JUST HAD SEX AND I NEED YOU TO TAKE ALL THESE FERTILIZED EMBRYOS I'VE MADE FAR AWAY FROM HERE SO THEY CAN GO FORTH YONDER UNTO THIS WORLD. (IN CONSIDERATION OF SAID SERVICES, YOU MAY DRINK NECTAR FROM MY SWEET FLOWER OR EAT MY SUGAR-SWEET FRUIT.) SOUND GOOD? GREAT, DONE DEAL.

But when something tries to take sugar without giving anything in return, the gloves come off. When a caterpillar, for example, chews, rips, and tears plant tissue, plants do a bunch of things in response. Electrical and chemical signals travel to the rest of the plant, alerting it to damage. Long, thin proteins inside the sieve tube called forisomes double or triple in width, partially blocking the tube. The cell starts producing a sugar called callose that also helps to plug up the tube.

But Mabel knows that this defensive dance is coming. So as soon as she confirms that the cell she's penetrated is a sieve tube, she spits out a different kind of saliva that pretty much stops the plant's defensive response in its tracks. Now she's basically set. She's suppressed the plant's sieve tube defense system, and because the tube is under pressure, she doesn't even have suck up the sap. She just opens or closes a valve in her head to control the flow.

But there is one more plant sap defense Mabel has to deal with: sugar. Specifically, the can-of-Coke–high or sometimes even Aunt Jemima–high concentration of sugar in sieve tube sap. As this incredibly concentrated syrup travels through Mabel's digestive tract, it encourages water out of her cells,[*] so much so that other cells, deeper in Mabel's gut, have to send their water to the front lines to replenish their fellow troops. Unfortunately, Mabel's gotta eat, so she keeps gulping, and this water loss continues. The more sap passes through Mabel and out her butt, the more water is "sucked" out of her body. Eventually, if she doesn't stop feeding on this plant syrup, Mabel will lose so much water to the sap that she'll dry out, shrivel up, and die.

Or at least she would . . . if she didn't have two elegant methods

....

[*] By a process called osmosis. If you want to see osmosis in action, dissolve a couple teaspoons of salt in a cup of water and then throw in a crisp romaine lettuce leaf. Come back in 20 minutes and you should find a very limp leaf—because all that salt leached out the water in the lettuce's cells. Same deal with Mabel, except instead of salt, sugar is the culprit, and instead of lettuce cells, Mabel's cells are the things in danger of shriveling up.

to deal with this water-loss problem. The first is the simplest: Every once in a while, Mabel might retract her stylet from the sugar superhighway, find some xylem—which carries water from the roots upward—and take a delicious draft of water to restore her dehydrated tissues. Second, Mabel has an enzyme in her gut that bonds sugar molecules together, which reduces the sap's ability to suck water out of Mabel's cells. All this is great for Mabel but terrible for the plant, because it means Mabel can feed for basically as long as she wants to.

Now let's take a moment and appreciate how insane this is. A creature smaller than your fingernail and lighter than a single four-inch-long strand of hair is able to

1. **worm its flexible-needle-mouth–face between individual plant cells, down to millimeters beneath the surface of a stem (or even tree bark in some cases).**

2. **find and penetrate plant cells transporting a 30 percent sugar solution at 100 to 200 pounds per square inch of pressure.**

3. **drink that plant sap almost undetected for basically as long as it wants without shriveling up and dying from the highly concentrated sugar solution sucking the water out of its body.**

There is no human analogy for this . . . but if there were, it would involve you building a hypodermic needle roughly the width of a toilet paper roll and the length of your left leg, somehow clamping that needle to your mouth, sneaking up behind a firefighter in the middle of spraying a house, stabbing the firefighter's hose with your toilet-paper-roll-size needle-face, attempting to control the resulting surge of water into your digestive tract, and doing all this without the firefighter noticing.

But let's get back to Mabel. She ain't done.

Mabel doesn't just sip on the sap. She chugs it. Why? For a surprisingly familiar reason: essential amino acids. Even if you don't know what these molecules look like, you've probably heard the term. Amino acids are the building blocks of protein, and there are roughly twenty different amino acids in all of nature. Your body—and the bodies of most animals, including Mabel—can produce about half of these, so you don't need to include them in your diet. As for the other half, you—and also Mabel—need to eat those; otherwise your body can't make the proteins it needs, and all manner of bad things happen to you. Plant sap happens to contain every kind of amino acid Mabel could possibly desire,* but in oh-so-minuscule amounts. So, to get enough essential amino acids, and also because the sap is under so much pressure that she doesn't have a choice, Mabel has to drink a buttload of sap.

And that means Mabel craps a lot.

Aphid poop is not like your poop. Chemically it's not all that different from sap; it's a clear and colorless sweet, syrupy liquid. You might already know it by a different name: honeydew. When Mabel was a wee lass, she could poop her entire body weight *every hour.* As an adult, Mabel poops about a milligram per hour. That doesn't sound like a lot, but remember that she weighs only *two* milligrams. Even if you were being force-fed like a foie gras goose and simultaneously had the worst case of diarrhea ever

....

* To be super-technical, the plant sap itself actually *doesn't* have every amino acid Mabel needs. But there are bacteria in Mabel's gut that convert amino acids in plant sap to the essential amino acids Mabel needs. Aphids have a microbiome, too, just like we do.

recorded, you would still be physically incapable of generating half your body weight of poo every hour.

All that is *one* aphid. When you start talking in terms of aphid *colonies*, it can be hard to wrap your head around the quantity of crap. In some forests, colonies can produce up to 110 pounds of dry honeydew per year on a single tree.* Depending on how dense the forest is and how many aphids there are, you could be talking about numbers in the hundreds-of-kilograms-of-honeydew-per-acre-per-year range.

But even after all this honeydew, Mabel still ain't done. Aphids have a complicated life cycle and reproductive strategy. In the winter Mabel can choose to bone a male aphid, producing eggs with a mix of her and the father's DNA. But in the summer, Mabel does *not* bone a male . . . but she *still* gives birth, this time to a live, fully formed, genetically identical copy of herself: a clone. And that baby clone is already pregnant with *another* clone when *she's* born. Scientists have a great name for this: telescoping generations.

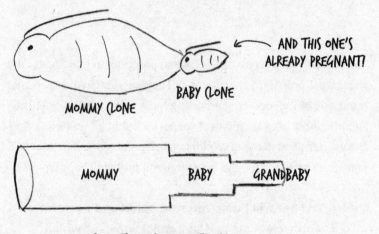

AND THIS ONE'S ALREADY PREGNANT!

BABY CLONE

MOMMY CLONE

MOMMY   BABY   GRANDBABY

SEE, IT LOOKS LIKE A TELESCOPE!

....

* In case you're wondering, "dry" honeydew means honeydew after the water has evaporated, which in turn means that the actual weight of poop exiting the colony's collective buttholes would have been much higher.

All this means that—assuming they don't get eaten by lady-bugs or other predators—there can be twenty generations of aphids in a single season.

So, to recap, plants are basically a gigantic all-you-can-eat aphid buffet. If Mabel and her kin like what's on the menu, they will:

1. **gorge themselves on sieve tube sap for days, depriving plants of their essential nutrient flow.**

2. **reproduce like it's their job.**

3. **coat everything below them in a sticky sugary mess, like a two-year-old with an ice cream cone.**

If you happened to be a member of, for example, the Tübatulabal people living in California thousands of years ago, you wouldn't have been all that bothered by any of that. In fact, you might have found a way to use it to your advantage . . .

The first people who made the critical observation that aphids shit sugar were probably . . . not people. They were *ants*. And even after hundreds of millions of years, some species of ants still get their sugar fix from aphids. Today, if you let an aphid (of species *P. cimiformis*) get its stylet into a plant, then put an ant (of species *T. semilaeve*) on the stalk, the following will probably happen:

1. **The ant will bump into the aphid and perform what scientists call "antennal waving," which looks roughly like fast-forwarded footage of a Pentecostal preacher faith-healing a parishioner.**

2. In response, the aphid will kick its rear legs, poop a droplet of honeydew, and point its honeydew-laden butthole at the ant. (Scientists call this "anal pointing.")

3. The ant will then receive this droplet of honeydew with reverence and begin to drink.

4. The ant will then "antennate"—i.e., feel up with its antennae—the aphid's butthole area, presumably to ensure that the flow of honeydew continues.

In return for the constant stream of honeydew, the ants protect the aphids from other predators. It's a classic symbiotic relationship.

It's all very well and good for *ants* to drink directly from the butts of aphids, but if you happened to one of the Tübatulabal, Owens Valley Paiute, Surprise Valley Paiute, Yavapai, Tohono O'odham, or a number of other Native American peoples a few hundred years ago, you had to be a little more creative. Watching the aphids carefully over the course of the summer, you probably noticed that, after a while, water evaporated from honeydew, leaving a coating of crystallized sugar on whatever poor plant they decided to infest—in California, usually a tule or reed or tall grass. And that was the basis of an ingenious processing method for something you could either call "honeydew spheres" or, somewhat more literally, "aphid shitballs." In the late summer or early fall, before the rains began, people would cut the stalks of long summer grasses that aphids had been feeding on, let those stalks fully dry in the hot sun, and then thresh—i.e., beat the hell out of—them with sticks over bearskin or deerskin. As the grasses were vigorously flayed, the honeydew would fall off the stalk and onto the animal skin. Then the honeydew could be gathered and shaped into little cakes or balls and eaten as is or warmed up by the fire.

By the way, Native Americans were (and still are) extremely sophisticated when it came to processing things from nature. Aphid candy is one example among many of things that you or I couldn't just wander out into a forest and make. Could you make baby diapers from lichen and exactly the right combination of algae and fungi? Or glue from sheep horns? Or gloves from mud hen skins?* No. You could not. Most of us wouldn't last five days alone in a national park, whereas a woman from the Tongva people survived entirely by herself on an island one-third the area of Washington, D.C., for *eighteen years*.

Today, food is evaluated based on how unsullied it appears. Heavenly food is ancient, organic, natural, and untouched by man. Hellish food is modern, industrial, and ultra-processed. But our actual history complicates these two categories. Where would aphid shitballs fall on this spectrum? I can just as easily imagine them shrink-wrapped and sold at gas stations as I can wrapped in brown paper and sold at Whole Foods. What about freeze-dried potatoes? Or detoxified cassava?

Maybe the road to hell—cobblestoned with Reese's, studded with Gushers, and sprinkled with Cheeto dust—isn't as new as it seems. Maybe where we are today is the result of very, very old trends. First, food detoxification: given the choice between dying and not dying, humans will exercise some ingenuity and effort to avoid death, whether by spending hours grating cassava tubers, freeze-drying potatoes, or a whole host of other techniques. Sec-

....
* You don't even know what a mud hen *is*. (It's a sort of duck-like creature.)

ond, preservation. Some people say that necessity is the mother of invention. That may be true, but I would add that laziness can also be a powerful motivator. Why risk the spoils of the hunt rotting, thus forcing you to hunt again? Much better to figure out how to *preserve* this dead thing, whether plant or animal, so you can spend more of your time lazing around camp. And, of course, preserving food is super helpful if you're trying to survive through the winter. Finally, food flavorification: once humans succeeded in eating foods and living to tell the tale, I can imagine them getting dissatisfied with the quantity of sweetness, saltiness, and fattiness available in the world around them and trying to concentrate those flavors or create new ones, whether by eating aphid poop, or more recently, breeding sugar beets and using the resulting sugar in every food product imaginable.*

The one thing our ancestors *didn't* do is worry about whether all this processed food would give them cancer. Why? Because, in their time, threats had . . . fur. Or exoskeletons and eight legs. Or they grew in the ground. Most of what was trying to kill you was alive.† Threats were not subtle, and they were not industrially produced. The chemicals we exposed ourselves to were "chosen" mostly by nature. Today, for most of the world, life is a lot easier and a lot less dangerous. Some threats are still furry and have eight eyes, but we are a lot less likely to be killed by a venomous spider or an infection and much more likely to be

....

* These trends aren't separate roads; they're more like lanes on the same highway. Preserving a food can also concentrate its flavors, like with jam. Detoxifying a food can also preserve it, like with the freeze-dried potatoes of the Aymara. Fermentation is particularly line blurring: it's hard, maybe even impossible, to ferment a food without changing its flavor, sometimes to the point where the exact same dead thing is considered a delicacy in one culture and a rotting pile of trash in another. (Google *surströmming* to see what I mean.) And, of course there are many other reasons besides detoxification, preservation, and flavorification that we processed stuff from nature; for example, we invented tea at least in part to get high on caffeine.

† Just like in modern-day Australia.

killed by heart disease or cancer. And as we've seen, ultra-processed foods have been linked to both of those diseases.

So, the argument goes, ultra-processed food is bad because it's modern, industrial, and unnatural. But what if, instead of arguing about whether a food or chemical is heavenly or hellish, we look at whether it's *optional*. How is that helpful? Well, first, a bit of background: we didn't invent the need to eat food. Or breathe air. Or drink water. But there are lots of chemical exposures we did invent: for example, inhaling an aerosol created by burning the leaves of a weed, also known as smoking. Or applying a white substance with a consistency halfway between butter and oil to our skin before going out in the sun, also known as sunscreen. Both of these are chemical exposures, and both are completely and utterly optional. So, to figure out if ultra-processed food is bad for you—and, by extension, to figure out which foods are good or bad—we're going to start by talking about some things that are *not* food.

First, we'll talk about smoking and vaping.

Then we'll look at sunscreen.

And finally we'll take what we learn from these two chemical exposures and see what that can teach us about the ultimate chemical exposure that is food.

PART II:
HOW BAD IS
BAD?

"PUT DOWN THAT CIGARETTE,
IT'S BAD FOR YOU."

—God

THE SMOKING GUN, OR WHAT CERTAINTY LOOKS LIKE

This chapter is about cigarettes, Spanish ribbed newts, exploding batteries, teeth, and xeroderma pigmentosum.

You probably know that smoking is bad for you, because your parents told you so. But how did *they* know? Probably because the surgeon general of the United States told *them* so in 1964. But how did Luther Terry know?

Not how you think.

The clearest way to demonstrate that smoking is bad for your health is to do a randomized controlled trial, much like we talked about in chapter one: take a bunch of people who don't already smoke, divide them up into two pretty much equal groups (on separate deserted islands), prevent one group from smoking, make the other group smoke, and check in on both groups yearly for the next fifty years.

This study has never been done. Why? Because it would be wildly expensive and a gigantic pain in the ass. But then again, they built the Burj Khalifa. The real reason this trial hasn't been done is ethics. Even back in the 1950s, people strongly suspected that smoking was bad for you, so no ethical researcher would have enrolled a nonsmoking patient in a trial that might require the patient to *start* smoking. Plus, most nonsmokers don't want to smoke. So it's hard to imagine them volunteering for a trial in which they might be required to do The One Thing They've Consciously Chosen Not to Do. For those reasons, a randomized controlled trial on smoking was never performed, and never will be.[*]

So how does science know that smoking is bad for you? Let us count the ways. First, we know that cigarette smoke contains at least seventy different molecules that can each cause cancer on their own. Remember formaldehyde, that super-promiscuous small molecule that will pretty much react with any biological molecule (chapter three)? Turns out, formaldehyde causes cancer in humans and is in cigarette smoke. So is benzene. And arsenic, which, in addition to being the poison of choice for a few hundred years back in the Middle Ages, is also a carcinogen at lower, not immediately deadly doses.

You might reasonably ask how we know that each of those seventy molecules causes cancer. For many of them, it's because a

....

[*] Unfortunately, lots of wildly unethical trials (unrelated to smoking) have already been run. For example, in the late 1940s, a doctor in the Public Health Service of the United States went to Guatemala and deliberately infected hundreds of people with gonorrhea and syphilis, sometimes by transferring gonorrhea-infected pus to the cervix of sex workers and then paying them to have sex with soldiers. Seriously. This. Actually. Happened. Years later, that same doctor participated in the infamous Tuskegee syphilis study, in which the U.S. Public Health Service knowingly withheld treatment from hundreds of black men with syphilis for twenty-five years because the government wanted to see what would happen if the disease ran its course untreated. After the predictable and highly justified uproar when Tuskegee became public knowledge, the government put formal regulations in place to safeguard against unethical research in humans.

particular trade (e.g., chimney sweeps in nineteenth-century London) was exposed to very high levels of a chemical (e.g., soot) and people in that trade developed absurdly high rates of cancer (e.g., scrotal cancer). Other chemicals, like arsenic, occur naturally in drinking water in certain parts of the world, and you end up seeing lots of cases of cancer there. Then there's animal experimentation. Every one of the seventy-plus chemicals in cigarette smoke has been independently given to almost every imaginable species of animal in the course of thousands of individual experiments by hundreds of scientists over the past fifty-plus years. All of those chemicals reliably cause cancer in at least one species.

Let's zoom in briefly and talk about one specific class of chemical in tobacco smoke called N-nitrosamines. These molecular mafiosi cause cancer in rainbow trout, zebrafish, medakas, guppies, platyfish-swordtail hybrids, Spanish ribbed newts, palmate newts, African clawed frogs, northern clawed frogs, grass frogs, ducks, chickens, grass parakeets, opossums, Algerian hedgehogs, tree shrews, European hamsters, Syrian golden hamsters, Chinese hamsters, migratory hamsters,* Dzungarian dwarf hamsters, gerbils, white-tailed rats, regular rats, mice, guinea pigs, minks, dogs, cats, rabbits, pigs, thick-tailed bush babies, capuchin monkeys, grass monkeys, patas monkeys, rhesus monkeys, and cynomolgus monkeys.

That's thirty-seven different species.

In addition to giving one chemical to a bunch of different species of animal, scientists also administered that chemical to one species in different ways. For example, let's look at one chemical in the N-nitrosamine family called NNK.

Scientists put NNK into rats' drinking water.

Result: lung cancer.

....
* Also known as the wanderlust hamster.

They injected it under rats' skin.

Result: lung cancer.

They inserted it into rats' stomachs via feeding tubes.

Result: lung cancer.

They swabbed it in the insides of rats' mouths.

Result: lung cancer.

They inserted it *directly into rats' bladders* via catheters.

Result: *still* lung cancer!

Not only did scientists test different species and routes of administration, they also tried varying the dose. That's kind of intuitive: if you increase the dose of a toxin and the symptoms get worse, that's a decent clue that the toxin might have something to do with those symptoms. A series of ten experiments done by scientists in at least three different institutions established what's called a "dose-response curve" but which I like to call a "how far up shit's creek you are" curve. Essentially, the scientists gave different groups of rats different doses of NNK and recorded what percentage of them developed lung cancer at each dose. For example, roughly 5 percent of rats given 0.034 milligrams per

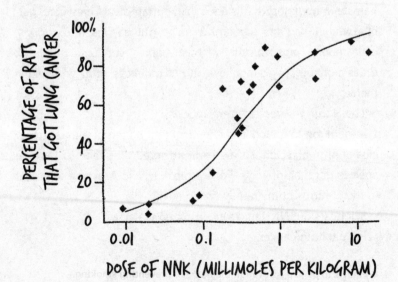

kilogram of body weight three times weekly for twenty weeks developed lung cancer, but when the dose was increased to .3 milligrams per kilogram of body weight, 50 percent of rats got lung cancer. At 10 milligrams per kilogram, roughly 90 percent did. (For reference, the dose of cyanide that kills roughly 50 percent of rats is about 5 milligrams per kilogram.)

As you can imagine, these experiments are a lot of work for scientists and a lot of cancer for rodents. In the roughly twenty-year period between 1978 and 1997, scientists published *eighty-eight* studies in which thousands of unlucky mice, rats, and hamsters were given NNK (and lucky ones were not). Animals given NNK developed markedly more cancer than the animals not given NNK. All these studies—and many more—show pretty convincingly that NNK and other N-nitrosamines are potent carcinogens in lots of different animals.

But hang on. Showing that there are known animal and human carcinogens in cigarette smoke—no matter how convincingly— doesn't actually prove that smoking cigarettes is bad for you. You could imagine Big Tobacco saying, *Sure, smoke has some chemicals in it, but it's only in contact with your lungs for half a second before you exhale. None of those chemicals actually stay inside the human body.*

Except they do, and we know they do in at least three different ways. First: the infamous black lung. Remember in high school when your teacher showed you that jet-black diseased-looking lung and told you it came from a smoker? It turns out that those "demonstration" lungs come from pigs, and since barnyard animals don't usually go through two packs a day for twenty years, those lungs are artificially stained brown or black.* So if you had X-ray vision and could see into a smoker's chest, it would *not* look like a coal mine. But if you were to compare an actual smoker's

....
* It hurts to be lied to, even in the name of preventing smoking.

lung with a nonsmoker's lung under the microscope, you would see lots of cells called macrophages in both lungs. These cells are part of your immune system and basically gobble up any foreign material—including smoke particles—to try and prevent it from doing damage. But in the smoker's lung, depending on how long the person had been smoking, the macrophages would look yellow, brown, or even black. This is because smoke particles are chemically difficult to break down, so macrophages store them in little compartments within themselves. Imagine your parents' basement: trash bags full of useless and dangerous crap that they can't throw away. Same idea. When enough of these particles accumulate, they become visible as little yellow or brown dots. The more you smoke, the more speckled your lungs become.

The second way we know cigarette chemicals get inside you is thanks to radioactive tracer studies, in which scientists use radioactive atoms to label certain molecules and then use a fancy Geiger counter to figure out how much radioactivity (which means how much of the labeled molecule) is in whatever organ they're looking at. There have been a ton of radioactive tracer studies done over the years, but one in particular stands out: in 2010, scientists published a study in which they radiolabeled nicotine in cigarettes, put some people in a radiation body scanner, and asked them to smoke *one single puff* of a cigarette containing radioactively labeled nicotine. Roughly twelve seconds after the puff, radioactivity was detectable in the subject's lungs; roughly twenty-two seconds post-puff, it was detectable in blood at the subject's wrist; and roughly fifty seconds post-puff, it was detectable in the subject's brain. This is pretty damn remarkable: it's probably the closest we'll get (for a while, anyway) to *seeing* a chemical spread out throughout the body as time passes.

The third way we know that chemicals in cigarettes get inside you is: pee. There have been tens, probably even hundreds, of what are called urinary metabolite biomarker studies, which is science-speak for "measuring specific chemicals in pee." But let's back up. You've heard the word *metabolism*, probably in the context of something like *Bruh, it's cold outside, my metabolism is slooooooow today*. But metabolism is much more than how quickly you burn your food; it's a spiderweb of chemical reactions that determine the fate of every molecule that gets into your body—food, drink, drug, or cigarette smoke. Your metabolism changes the molecules in cigarette smoke to make them dissolve better in water. This helps your body pee them out. Once they're in your pee, scientists can measure them. The tricky part is that there are so many chemicals in cigarette smoke and so many metabolic reactions that it can be tough to figure out which chemicals came from cigarettes and which ones came from food, drink, other drugs, or the environment around you. There have been hundreds of studies comparing smokers with nonsmokers to try to clear up this mystery. Eventually scientists zeroed in on a hit list of eight possible biomarkers that were all chemically related to carcinogens in cigarette smoke. Then, in 2009, a group of scientists published a study in which they:

1. **found seventeen smokers.**
2. **measured levels of those eight chemicals in their blood.**
3. **required the smokers to quit smoking.**
4. **kept measuring the eight chemicals every couple of weeks for two months.**

Within three days of quitting, levels of five of the eight biomarkers went down by 80 percent or more. Levels of another went down by roughly 50 percent. And the seventh biomarker took about twelve days to reach 80 percent reduction. Only one of the eight didn't show any reduction after quitting. This experiment is particularly convincing because you're not comparing two different people; you're comparing the *same person* as a smoker and then as a nonsmoker.

So scientists have established beyond a reasonable doubt that cigarette smoke contains carcinogens and that smoking brings those carcinogens into your body. This might seem like a lot of evidence—and it is—but it's not enough to prove that smoking cigarettes causes lung cancer. All we've shown so far is that smoking cigarettes brings carcinogens into your body. What happens once they're inside you?

Answering this question requires figuring out exactly what happens to each of the seventy-plus carcinogens in cigarette smoke once they get inside you: their "metabolic fate." It turns out that carcinogens are not usually carcinogenic in their initial form. But as they pass through our metabolic machinery (specifically a protein with the Terminator-like name of cytochrome P450) they are converted—for an infinitesimally short time—to an activated form, meaning their chemical reactivity is turned up to an 11. Most of the time, they're safely deactivated and peed out, but occasionally an activated molecule can slip away and form a chemical bond with something else in the cell, which, even more occasionally, is our old friend DNA. This general pathway—carcinogen gets into cells, gets activated by cytochrome P450, and then binds to DNA— has been tested in repeated experiments for hundreds of carcinogens over the past seventy-plus years, both with carcinogens found in cigarettes and those found outside of cigarettes.

Thus we have another link in the chain: carcinogens in cigarette smoke bind to DNA. But, believe it or not, chemicals binding to DNA *still doesn't prove* that they cause cancer. We have to figure out what happens to that chemically altered DNA.

When chemicals bind to DNA in ways your body doesn't expect (the way cigarette smoke carcinogens do), your body deals with the damaged DNA the way you try to deal with a damaged computer: by trying to fix the damn thing. In the best-case scenario, a cell successfully repairs its DNA, and you go on with life as if nothing happened. Occasionally, though, the damage is unfixable or the repair effort fails. In that case, the cell goes, *I'M OUT,* and kills itself.* This might seem bad, but it's not the worst thing that could happen. There are a couple scenarios that are worse. The cell could fix the damage but do a bad job. Or it could fail to detect the damage before starting to copy its DNA . . . and copy it wrong. Either way, the result is a mutation.

You've probably heard of DNA mutations. A mutation is a change in the genetic code, the blueprint the cell uses to live its life. If our logical chain of reasoning so far is correct, we would expect smokers to have *more* mutations in their DNA than nonsmokers, because they've had more chemicals bind to their DNA than nonsmokers. And they do. There haven't been quite as many studies done to support this part of the logical chain, because large-scale DNA sequencing has only recently gotten cheap enough to do routinely. But in one striking example, researchers surgically removed a lung tumor from a fifty-one-year-old man who had been smoking twenty-five cigarettes per day for fifteen years and found more than 50,000 mutations compared to a nonsmoker's genome. Other studies haven't shown quite as dramatic a difference, but they consistently show that smokers have many more mutations than nonsmokers.

We're still not done. Scientists have shown pretty convincingly that smokers have more mutations in their DNA, but how do we know that mutations cause cancer?

Between 1938 and 2017, the U.S. government appropriated

....

* Suicide isn't quite the right analogy here. What the cell actually does is kill itself, then plan its own funeral, then plan its own estate sale, then chop itself up into little pieces so it can be recycled into other cells. Occasionally, the cell won't notice how damaged it is. In that case, a neighboring immune cell takes it aside and convinces it to put its affairs in order and say goodbye.

almost $130 billion to the National Cancer Institute. Today, NCI spends roughly $5 billion per year on cancer research, making cancer the number one destination of our research dollars. A lot of this money goes toward figuring out what causes cancer, and the consensus answer is that DNA mutations cause, or help drive the growth of, lots of different cancers.[*] Let's look at two pieces of evidence (out of many) that support this idea.

One comes from a completely different field. There's a rare human disease called xeroderma pigmentosum, or XP, which sounds like a Harry Potter spell but is actually a devastating disease. People with XP are *extremely* sensitive to the sun. They get badly sunburned within a few minutes of being in full sun, develop freckles pretty much everywhere they don't cover up, their eyes go red, and those under twenty have rates of skin cancer about 1,000,000 percent higher than normal. That's not a typo. One million percent higher. In 1968 scientists discovered that XP is caused by inherited mutations in a few key genes your body needs to repair its DNA. This fits in very nicely with the theory that DNA mutations can cause cancer: if your body is bad at fixing its DNA, DNA damage would become mutations much more frequently, and that would explain the incredibly high rates of cancer among people with XP.

Another piece of evidence is more related to smoking. Scientists recently sequenced thousands of genes from 188 lung tumors and found that the two most commonly mutated genes were *KRAS* and *TP53*, both of which we know—thanks to the $130 billion NCI has spent on cancer research—are involved in pushing cells to grow faster (*KRAS*) and preventing them from shutting themselves down if they get out of control (*TP53*). Both of these behaviors are classic markers of cancer cells. So that's an excellent clue, but to really

....

[*] As with all consensus answers, there are scientists who disagree. But that's another book.

prove that mutations in those genes *cause* cancer, you have to actually mutate them and see what happens. Amazingly, we have the technology to insert mutant copies of *KRAS* and *TP53* into human egg cells and see if the resulting humans get lung cancer . . . but that might be one of the cruelest studies ever conceived. Instead, scientists mutated both those genes in fifty-six mice. *Every single mouse* got lung cancer. In nineteen mice (34 percent), the cancer metastasized. By comparison, only 5 percent of mice with *just KRAS* mutated had metastatic cancer.

Are you ready for a fun twist? Here's a wild but true statistic: 10 to 20 percent of smokers get lung cancer. You can view this statistic in two ways: either *HOLY SHIT, ONE IN SIX SMOKERS GETS LUNG CANCER, A DISEASE YOU'RE ALMOST CERTAINLY NOT GOING TO GET IF YOU DON'T SMOKE* or *HOLY SHIT, HOW THE HELL DOES REGULARLY INHALING SEVENTY CARCINOGENS NOT GIVE EVERY SINGLE PERSON WHO DOES IT LUNG CANCER?* I'm in the latter camp: my feeling is that if cigarettes so clearly cause lung cancer, it's kind of amazing that *every* smoker doesn't get it. However you see it, you might think this piece of information throws a major wrench into the smoking/lung cancer hypothesis. But how wrench-y is the wrench? Let's revisit the last link in the chain: *DNA mutations in certain genes can lead to cancer.* Your genome is 3 billion pairs of letters long, and most of those letters don't actually code for anything. Assuming smoking causes random mutations throughout your genome, the probability that any single mutation will happen in a gene linked to cancer like *TP53* or *KRAS* is roughly one in a million (per cell). So it's entirely possible that a smoker could go their entire life and never get a mutation in either of these genes, in the same way that it's possible to drive tipsy your entire life and never get into an accident.

That's one way to explain the wrench. We could also revisit the penultimate link in the chain: *DNA damage, if not properly fixed, can lead to mutations.* What if certain people are just better at fixing their own DNA than others? We've already seen that certain people who are particularly *bad* at fixing their DNA (those with xeroderma pigmentosum) end up with insanely high rates of skin cancer if they aren't super-careful to avoid ultraviolet light their entire lives. It's conceivable that certain other people are particularly *good* at fixing their DNA. In this group, you'd guess that smoking would cause just as much DNA damage as in anyone else, but it would be repaired more quickly and with fewer mistakes. So most DNA damage would *not* lead to mutations, and these people could smoke their whole lives and never get lung cancer.

There's a third way to explain the wrench. In addition to lung cancer, smoking also causes a smorgasbord of other diseases, including cardiovascular disease and stroke. So you could die of a heart attack way before smoking would have given you lung cancer.

Are you ready for another fun twist?

All the experiments we've looked at in the previous pages were done *after* the surgeon general issued the 1964 report. And yet, knowing very little about the exact chemical mechanism of *how* smoking causes cancer, the authors of the report wrote: "Cigarette smoking is *causally* related to lung cancer in men" (emphasis mine) and "The risk of developing lung cancer increases with duration of smoking and the number of cigarettes smoked per day, and is diminished by discontinuing smoking."

They did not write "seems to be related to" or "may cause" or "might influence" or "could potentially be a contributing factor to the genesis of the existence of" lung cancer. They just came right out and told the nation that smoking *is* a cause of lung cancer.

How could they have been so sure? Remember, there had never been one single solitary randomized controlled trial on the long-term health effects of smoking.

First, you should know three pieces of background information.

Piece one: In the early 1960s, about 40 percent of Americans smoked, and each smoker blew through over 4,000 cigarettes per year, on average. That's roughly half a pack per day.

Piece two: Before the early 1900s, lung cancer was an incredibly rare disease—so rare that, in 1898, one single solitary PhD student wrote one article that reviewed *all the lung cancer cases in the world*: 140 cases. Throughout the twentieth century, there was an incredible rise in lung cancer cases, which paralleled—with a three-decade delay—the rise in cigarette sales.

Piece three: About 60 percent of Americans *didn't* smoke, so there were plenty of nonsmokers available to compare against the smokers. Lots of people were ready and waiting to become science.

These three things frame what may well be the most ambitious knowledge-gathering exercise humanity has ever performed. In the late 1950s and early 1960s, well over a million people were enrolled in studies on smoking: their every ailment, condition, disease, morbidity, and dysfunction observed, classified, verified, and recorded, from the moment they were enrolled to the moment they died (or didn't). Some studies were relatively small and short; others included half a million people and are still underway to this day, more than fifty years later. All of them were "prospective cohort" studies.

As we saw in chapter one, in these studies, you recruit a bunch of people, put them through a medical exam, get them to tell you whether they smoke and how much, and then follow everyone for years and see which group has more lung cancer (or heart disease, or death, or whatever outcome you're interested in). The concept is similar to a randomized controlled trial, except that you don't require people to smoke (or not smoke). You just find people and record whether they are already smoking (or not).

The authors of the surgeon general's report relied on data from seven cohort studies to investigate whether there was a link between smoking and lung cancer. In one British study, every participant was a doctor. (Back in 1964, a lot of doctors smoked.) In another, every participant was a veteran. The largest study—448,000 participants—was of American men in twenty-five states. Some studies had been operational for just five years. Some had been running for over twelve. All told, the studies had enrolled over a million people in England, Canada, and the United States.

The results were stunning. On average across all studies, smokers were about *eleven times* as likely to die of lung cancer than nonsmokers. "Eleven times as likely" means 1,100 percent—one thousand one hundred percent—as likely. If you took a nonsmoker's risk of dying from lung cancer, doubled it, doubled it again, and doubled it a third time, it would *still* be lower than a smoker's risk of dying from lung cancer.

Let's take the evidence train to the next logical station. If cigarettes kill you, you would expect *more* cigarettes to kill you *more*.[*] Four of the seven cohort studies tracked how many cigarettes participants smoked. In every single study, the risk of dying from lung cancer increased steeply with increasing numbers of cigarettes smoked. Likewise, you'd expect that people who inhaled more deeply would get a higher lifetime dose of cigarette smoke. And indeed, people who inhaled "deeply" had a 120 percent greater risk of dying than nonsmokers.

No matter how the authors sliced and diced the numbers, the conclusion from all the cohort studies was the same: people who

....

* This is true for cigarettes but may not be true for the other myriad things that are trying to kill you. When it comes to acutely toxic poisons like cyanide, for example, any amount over a certain threshold will kill you. The relationship between *dose* and *effect* can be complicated.

smoked were much, *much* more likely to die from lung cancer than people who didn't smoke; and they were also more likely to die from other diseases than people who didn't smoke.

But did that mean that cigarette smoking *caused* lung cancer?

The tobacco industry had been insisting for years that the answer was "not necessarily," using arguments like this: *Yes, there is a parallel between the sale of cigarettes and lung cancer, but there is also a parallel between the sale of silk stockings and cancer of the lung* and also that *simply because one finds bullfrogs after a rain does not mean that it rained bullfrogs.* Allow me to add my own metaphor to the mix: just because a warm pie is sitting on your kitchen counter doesn't mean your mom came over unannounced and baked it. The point of all these metaphors is this: just because two things are associated doesn't necessarily mean that one thing caused the other. There may very well be alternate explanations. Silk got more fashionable. Bullfrogs converge after a rain to eat worms. Little Red Riding Hood could have brought the pie over from her place, you could have ordered it off Seamless, it could have been reheated from two days ago, or aliens could have baked it and teleported it inside your house. Now there will silk-clad bullfrogs eating a pie baked by aliens in your dreams. You're welcome.

Anyway, lots of people debated alternate explanations other than smoking for the incredible rise in lung cancer: Were doctors simply better at diagnosing it? Could it be car exhaust or the paving of roads, both of which rapidly increased alongside cigarette smoking? What about industrial pollution? Or maybe it wasn't related to chemicals in the environment at all; maybe there was some gene that caused both a craving for cigarettes and lung cancer. So we come back to our original question: How were the authors of the surgeon general's report so sure that smoking was *causing* lung cancer?

It wasn't because they understood the *how.* Yes, there had been some animal experiments, the most notable of which involved condensing cigarette smoke and painting it on mouse skin, which resulted in skin cancer.[*] And scientists had also identified a small number of likely or known carcinogens in cigarette smoke. But the crux of the authors' reasoning was mostly based on the large prospective cohort studies that showed four things:

1. **Lung cancer happened after—not before—smoking.**

2. **The vast majority of lung cancers happened in smokers.**

3. **This link was found in a variety of different populations.**

4. **The increase in risk was huge and got higher the more you smoked or the deeper you inhaled.**

And there's one other thing to consider: the day-to-day experience of someone with lung cancer. It's not an easy cancer. You don't just get to chill out in a hospital, have a minor procedure, joke around with the nurses, and then go home to your cancer-free life. Even today, with all the advances of modern medicine, your chance of still being alive five years after a lung cancer diagnosis is about 19 percent. So, as the committee was considering whether to definitively say that smoking caused lung cancer, they must have had these considerations in the backs of their minds—and they knew that lung cancer deaths were skyrocketing. In 1898, it was a medical oddity. By 1964, it was killing more

....

* Somewhat amazingly, it would take researchers another fifty years to show that *inhaling* cigarette smoke causes lung cancer in mice. Turns out lab animals don't like to smoke. You pretty much have to put them in a chamber and fill it with cigarette smoke . . . kinda gruesome.

than 50,000 Americans every year. Today, that number is over 140,000. (Globally, lung cancer killed almost 1.8 million people in 2018.) So even though there wasn't the incredible breadth of mechanistic evidence linking smoking to lung cancer that we have today, there was a large body of observational evidence, no plausible alternative explanation, and a catastrophic potential downside to *not* speaking out. This was enough for the committee and the surgeon general to put a stake in the ground and announce definitively that smoking causes lung cancer.

Two scientists can look at the same data, experiment, or theory and have legitimate differences of opinion. One scientist's "obvious conclusion" is another's "spurious assertion." And everybody's threshold for truth—the mental line an idea must cross before it's accepted it as fact—is different. But almost every scientist looking at the full set of data on smoking would conclude that smoking causes lung cancer. Even though there are no randomized controlled trials on humans, there are thousands of experiments that together support a logical chain of events:

CIGARETTES CONTAIN CARCINOGENS

THOSE CARCINOGENS GET INSIDE YOUR BODY

ONCE THERE, THEY CHEMICALLY REACT WITH DNA

THIS SCREWS WITH IMPORTANT PROCESSES LIKE COPYING DNA, SO YOUR BODY TRIES TO FIX THE DAMAGE

OCCASIONALLY THIS RESULTS IN A MUTATION

MUTATIONS ACCUMULATE IN YOUR DNA

ENOUGH MUTATIONS IN GENES THAT CONTROL CELL GROWTH CAN PUT CELLS ON THE PATH TO CANCER.*

....

* And DNA isn't even the whole story. Cigarette smoke also contains chemicals that don't mutate DNA, but *do* push cells along the path to cancer. But that's another book.

Plus, more than a *million* people were enrolled in multiple long-term observational studies, and no matter where the study was done or who the participants were, the increase in lung cancer risk among smokers was stratospherically high and increased the more they smoked.

All the people involved—the participants in all these studies, the surgeon general's committee members, the scientists puzzling out the mechanism—and all the animals who died in service of the cause did something pretty frickin' incredible. They built a bridge from the Land of Not Knowing, shrouded in mist and shadow, to the Land of Almost Certainly Knowing. This bridge is built from thousands of experiments and studies, each a dense brick, supporting and being supported by others, without holes between them, together spanning the full width of the Gulf of Ignorance. We haven't talked about every single brick—not by a long shot. But hopefully all the bricks we *have* talked about, and the logical mortar that binds them together, give you a reasonable idea of what it looks like when a group of scientists well and truly *know* something.

Unfortunately, as we've seen before, scientists are really bad at naming things, and this is no exception. Instead of something cool like BRIDGE OF TRUTH, scientists decided to name these bridges "theories." As you might already know, the word *theory* means something very different in science than it does in English.

In English, a theory is usually a flimsy explanatory thought generated by an observation or two. For example: you wear a red shirt and win a golf tournament. Theory: red shirts make you better at golf.

In science, a theory is a solid and well-constructed Bridge of Truth. Gravity. Atoms. Evolution. All these are *scientific* theories: unlikely to collapse if you wear a red shirt and lose.

I'm not a fan of the word *theory*, for two reasons:

1. **It has two exactly opposing meanings in science vs. English (solid vs. flimsy), and**

2. **the English definition has already won.**

Let's not sugarcoat this last point. To most people, the word *theory* means "crazy idea that I just pulled out of my butt." And that's why a sentence like "scientists have developed a theory that smoking causes lung cancer" sounds so . . . unimpressive. We've all got that stubborn English definition kicking around in our stubborn brains.

But in the also-stubborn real world, scientists *know* that smoking causes lung cancer.

———

Despite the now utterly overwhelming evidence that smoking causes cancer—as well as heart disease and other things you definitely don't want—the tobacco industry is doing spectacularly well. Why? Basically, because it managed to export cigarettes from England and the U.S. to the rest of the world.[*] But as any good Stock Photo Businessperson will tell you, diversifying your integrations will synergize your corpuscles. In other words, it's better to not have all your eggs in one basket. For a long time Big Tobacco's one basket was the cigarette. And even though the cigarette is generating loads of cash outside the U.S., it sure would be nice (for Big Tobacco) if there were an alternate nicotine delivery device to help diversify the corpuscles here at home.

Enter the electronic cigarette, also known as the e-cigarette. We know a lot less about the chemistry and health effects of e-cigarette smoke (also called vapor) than we do about regular cigarette smoke,

....

[*] For this reason, and because of the long delay between starting to smoke and maybe getting lung cancer, cigarettes will actually kill more people this century than they did in the last.

but let's wade out into the evidence and see what we can see. The most dramatic difference between regular cigarettes and e-cigarettes actually has nothing to do with smoke. Regular cigarettes are powered by the energy released when tobacco burns. E-cigarettes, on the other hand, are powered by lithium ion batteries, which means that they sometimes decide to catch fire or explode, seemingly by themselves. This produces some truly horrific injuries.

In one case, an e-cigarette exploded while in the mouth of an eighteen-year-old, breaking one of his front teeth at the gum line, smashing another all the way up into his gum, and blowing a third clean out of its socket. In another case, a twenty-year-old man's e-cigarette suddenly exploded, shooting the mouthpiece into his face with so much force that it broke the right bridge of his nose, splintering the bone into pieces. And because apparently that wasn't enough, the battery shot out in the other direction and started a fire—adding arson to injury. In a third case, a twenty-six-year-old man was testing an experimental model when it failed catastrophically, spraying shrapnel into the man's chest and left shoulder. Remember that time Dick Cheney "accidentally" shot his hunting partner in the face? Yeah, this looked like that. (The doctors treating him called his skin injuries "shotgun-like.") Finally, there have been many cases of e-cigarettes exploding in people's pockets, causing burns and other havoc to their thighs. I've read at least two cases in which people ended up with second-degree burns or other damage to their woo-hoo parts. E-cig explosions are, of course, incredibly rare. But the damage shows just how much energy is packed into those little batteries.

The less dramatic but more obvious difference between regular cigarettes and e-cigs is that regular cigarettes have *many* more chemicals in them than e-cigarette liquid (a.k.a. e-liquid or e-juice). At first blush, this is counterintuitive: after all, regular cigarettes are basically just dried-up tobacco leaves rolled up in a piece of paper with a filter stuck at one end. If you count only the ingredients that get used up during smoking, that's a grand total of . . . two. In the opposite corner, a recent analysis of e-cigarette liquid suggests that despite having just three or four ingredients listed on the label, there could be sixty or more different chemicals floating around in there. But before you fall off your chair, remember that "tobacco" is not a single ingredient: each leaf was once a living thing made of countless cells, each carrying DNA, proteins, sugars, and a cacophony of other chemicals the plant made before it was picked, washed, cured, cut, and stuffed into a rolled piece of paper. So what looks like just two ingredients is actually many more: at last count, there were about 5,700 different chemicals identified in tobacco (not including additives), and the scientists who wrote the book on tobacco chemicals estimate that there are "literally tens of thousands" more yet to be identified.

But there's one chemical that looms large. You could say there's

One chemical to rule them all, one chemical to find them.
One chemical to bring them all, and in the smoker bind them.

Nicotine.

Nicotine is the main reason smokers keep smoking; it's what makes cigarettes addictive. It's also the whole point of e-cigarettes: they were invented to deliver nicotine without the side dose of cancer. In addition to being addictive, nicotine is also poisonous the old-fashioned way: if you get too much of it inside you, you'll die. If it makes you feel any better, we're not nicotine's intended target; tobacco plants make it to kill insects that would otherwise eat them. In other words, nicotine is a natural pesticide; in fact, it was extracted from tobacco and used as a pesticide as far back as the seventeenth century. And tobacco plants evolved a pretty potent poison: even the most conservative estimates put the lethal oral dose at around 10 milligrams per kilogram of body weight.

That means that a 30-milliter bottle of high-nicotine e-liquid contains enough nicotine to kill a fully grown adult, and *more* than enough to kill a toddler. You would have to eat 83 cigarettes—or smoke 603—to get the same amount of nicotine that's in that 30-milliliter bottle. Also, while cigarettes taste like . . . cigarettes, e-liquid can taste like "Birthday Cake," "Fruit Loops," or any of thousands of other flavors designed to make kids want to eat them.[*] So if you vape, please keep your e-juice away from kids: some of those bottles are literally candy-flavored poison.

Finally, we arrive at the least obvious—and most important—difference between regular and e-cigarettes: the smoke.

To understand the difference between "vape" and "smoke," we have to go back to the first difference. A regular cigarette is powered by a combustion reaction, which means smoking it is somewhat like sucking on the tip of a tiny campfire. If you took high school chemistry, you were probably taught that combustion looks like this:

$$\text{A NICE SIMPLE HYDROCARBON (LIKE METHANE)} + \text{OXYGEN} \rightarrow \text{CARBON DIOXIDE} + \text{WATER}$$

But there was also this thing called *in*complete combustion, which looked like this:

$$\text{A NICE SIMPLE HYDROCARBON (LIKE METHANE)} + \text{NOT ENOUGH OXYGEN} \rightarrow \text{CARBON MONOXIDE} + \text{WATER} + \text{CARBON}$$

....

[*] And it's not just eating. Some e-liquids come in bottles that make them look like prescription eyedrops. So . . . don't store them in your medicine cabinet right next to your eyedrops. Yes, this has happened. And, yes, it caused a woman to accidentally put e-liquid in her eye.

If a cigarette was (a) a simple hydrocarbon and (b) combusted completely, the only two products of the reaction would be carbon dioxide (a gas) and water (also a gas, since the burning happens at high temperatures). Cigarettes would disappear into thin air as you smoked them. Obviously, this does not happen. Cigarettes are chemically incredibly complex mixtures, and they do not combust completely. So instead of a nice simple reaction, the best we can do is something like this:

BURNING THING MADE UP OF THOUSANDS OF CHEMICALS + NOT ENOUGH OXYGEN → GIGANTIC CHEMICAL CLUSTERWHOOPS OF THOUSANDS OF CHEMICALS

Cigarette smoke is a wildly complex chemical cocktail. What about e-cigs?

If lighting up a regular cigarette is like sucking on the tip of a tiny campfire, then vaping is like a cross between huffing a can of hairspray and a Glade PlugIn. An e-cigarette is not powered by a combustion reaction; instead, a metal coil heats up the e-liquid to about 302°F to 662°F (150°C to 350°C—kind of like a plug-in air freshener), which generates a mist (kind of like what comes out of a can of hairspray). Regular cigarettes burn much hotter: 1,472°F (800°C) or higher. Because the temperature is much lower in an e-cigarette, and because e-liquid is chemically much simpler than tobacco, the number of chemicals in e-cig vape is almost certainly much lower than in regular cigarette smoke. Why *almost* certainly? Because e-cigarettes haven't been around for that long, and it can take a while to try and figure out what's in something. In 1960, fewer than five hundred chemicals had been identified in tobacco and cigarette smoke. (Those numbers have risen fairly steadily to more than 10 times the number today.) My point is: there is probably more to discover in e-cigarette mist—and we're starting to do exactly that.

Speaking of "mist," I have to hand it to whoever came up with that term (and "vaping"), because those words make it sound like you're sipping on a harmless fluffy cloud of water vapor . . . which you're not. Even though there's no combustion reaction as in a regular cigarette, the vaporizer gets hot enough to do some chemistry. For example, the heat can break down two of the most commonly used chemicals in e-liquid, propylene glycol and glycerin (a.k.a. glycerol), forming formaldehyde, acetaldehyde, and acrolein. You definitely don't want to be breathing in buckets of formaldehyde (remember chapter three?), and the same goes for the other two musketeers in this particular trio. All three of these guys have been reliably detected in many different brands of e-cigarette mist (though all at lower levels than in cigarette smoke), along with roughly eighty other chemicals along for the ride in the e-liquid or produced by its vaporization.

Let's pause here and take a quick detour into the world of chemical accounting.

As many people have already noted, 80 chemicals is a lot fewer than 5,700 chemicals. And if you're a vaper, you've probably seen a sentence in an ad that goes something like: "E-cigarettes don't have nearly as many chemicals as cigarettes do, so they're not as bad for you." And you've definitely seen at least one anti-smoking ad that says something like: "There's a toxic mix of more than 7,000 chemicals in every puff." The implication is clear: the more chemicals something contains, the worse for you it is.

This is, in my opinion, wildly illogical.

You know what else has thousands of chemicals in it? Iceberg lettuce. So does chicken. And lima beans. By contrast, cyanide has exactly one chemical in it. It *is* one chemical, and a very simple one at that—and it's deadly. The *number* of chemicals in something is the least useful piece of information you could be given. It is, like that Instagram post of your friend in the gym, marketing bullshit. It tells you nothing about *what those chemicals do in your body*, or even *how much of each chemical is in the thing*. Cigarettes cause lung cancer, but it's not because there is some chemical accountant who decided long ago that anything with more than thirty-seven chemicals in it must be toxic. Cigarettes cause lung cancer because of *which* chemicals are in smoke and *how much* of them there are, not how many.

So, for the chemicals we know are toxic, what's the difference between regular cigarettes and e-cigarettes?

From the handful or so experiments that have been done so far, it seems like e-cigarette mist has fewer known toxins than regular cigarette smoke, and the ones it *does* have are present at lower levels. For example, the amount of formaldehyde in e-cigarette mist is roughly one-tenth the amount in regular cigarette smoke, and the amount of NNK (the potent lung carcinogen) in e-cigarette mist is roughly one-fortieth the amount in regular cigarette smoke.

So let's break out the champagne-flavored e-juice, amirite?

Not quite yet.

Here's where we get into the two sides of the vape debate.

The Optimists: *E-cigs have way less toxic shit in them than regular cigs, so they're better for you!*

The Cautious Folk: *Just because they're better for you doesn't mean they're not bad for you. You're still inhaling an aerosol with a bunch of known toxins in it.*

I have to say that I find the Cautious Folk's argument slightly more convincing. Getting shot with a .22 is certainly healthier than getting shot with a .357, but that doesn't mean the .22 is healthy. Comparing e-cigarettes to regular cigarettes makes the e-cigarettes look pretty damn good, but that's mostly because regular cigarettes are *so* bad. It's entirely possible that e-cigs are much healthier for you than regular cigarettes *and also* increase your risk of lung cancer or other diseases. The more informative comparison is exactly the same one that was done for regular cigarettes: How bad are e-cigarettes for you compared to . . . nothing? We can reasonably guess that they're worse for you than nothing, but as for how much worse—we don't really know the answer to that yet. A few studies are starting to emerge, like early spring flowers poking through the snow, but e-cigs haven't been around that long, so the type of large long-term prospective cohort studies (like the ones scientists did on regular cigarettes) are still under way.

But there's yet *another* thing to consider: the ramp theory. Let me explain. If you want to quit smoking but keep the nicotine, there are a few different options available: the patch, gum, lozenges, inhalers, etc. But none of these replicate everything that goes along with smoking: lighting up, inhaling, the quick nicotine hit, pairing cigarettes with coffee, smoke breaks—in other words, the whole ritual. The pharmacist who invented the modern e-cigarette, Hon Lik, wanted to develop something that *both* delivered nicotine and also preserved the ritual, because he felt that was the best way to switch from regular cigarettes to something less harmful. And I don't blame him: this makes perfect sense. If you want smokers to quit, it's easier to guide them down a ramp than throw them off a cliff. But the problem with ramps is they can work both ways: you can imagine a scenario in which people who have never smoked in their lives might take up vaping, adopt all the smoking rituals, and eventually start smoking regular cigarettes.

This type of theorizing gets complicated fast, but the basic point is that the health impact of e-cigarettes doesn't *just* depend on how bad they are for you; it also depends on how vaping influences smoking—both starting and quitting.

So, tl;dr.

Smoking: we absolutely know it's terrible for you, and we know *how* terrible it is for you.

Vaping: we don't know exactly how bad it is for you, but we do know it's not *good* for you. That said, if you have to choose between smoking and vaping, all available evidence points strongly to vaping. It may help you quit, and it seems like it's not as bad for you as smoking. But if you're choosing between vaping and *nothing*, all available evidence points strongly to *nothing*, for three reasons. Reason one: vaping is almost certainly worse than good ol' plain air. Reason two: vaping can be an on-ramp to the thing we know is terrible for you—smoking. Reason three: the vape liquid could be contaminated.

If the tl;dr was tl and you dr, just remember this:

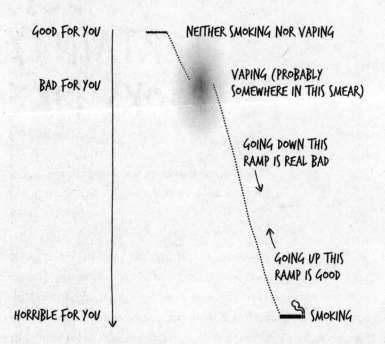

And if you do vape, keep that e-cig far away from your junk.

CHAPTER FIVE

SUNBURNT TO A CRISP, OR WHAT LESS CERTAINTY LOOKS LIKE

This chapter is about sunscreen, vitamin D, our genetic code, I Can't Believe It's Not Butter!, and coral reefs.

In 2012 a seventy-seven-year-old English woman left England and went on vacation to the South of France. One day she fell asleep out in the sun. She was wearing a patch containing the opioid fentanyl, which had been prescribed to her for back pain. Patches work by putting the drug in close contact with your skin; it slowly enters your body and makes its way into your bloodstream. It's a simple and elegant drug delivery system. Unfortunately, when your skin gets warmer—as it would in the sun—the amount of fentanyl (or any drug) that can diffuse into your body increases. If you've been reading anything about the opioid epidemic, you know what too much fentanyl can do.

The woman slipped into a coma.

Normally, when you fall asleep in the sun, your body eventually senses that you're getting too hot and wakes you up. Depending on how white you are, you might wake up with the unpleasant sensation that your skin has been deep fried (and not for the sake of preservation). You might even blister or peel over the next couple days. This is, of course, a sunburn, and—depending on how light your skin is and where you are in the world—you can develop one in as little as a few minutes.

The woman was in a coma under the hot French sun for *six hours*.

By the time the ambulance arrived, the woman had developed the worst case of sunburn that I could find in the medical record. Her sunburn was so severe it looked like a *fire* burn: charred black lines of literally cooked skin snaking across her abdomen and legs, and—even more awful—certain areas of pale white leathery burned *fat*, meaning that the sun had burned through all three layers of skin (about 2 millimeters deep) and then roasted the fat beneath. Her burns were so bad that, once out of the coma, she had to be treated in a specialized burn unit.

How did the sun do this? Let's consider the amount of energy involved. The sun is a colossal, gargantuan, ginormous, prodigiously powerful energy gun. If you took the amount of energy released by the nuclear bomb detonated over Nagasaki, Japan, and multiplied that by a thousand, you'd get the amount of energy that hits the Earth from the sun . . . *in one second*. What makes this case so surprising isn't so much that the sun *can* cook you; it's how rarely it happens. And for that we have our bodies to thank. They instinctively understand that too much sunlight is bad, and they have two extremely sophisticated ways of telling you so:

1. I am hot. (Translation: *Get inside now! Or find some damn shade!*)

2. I have sunburn. (Translation: *Feel my wrath, human; this is your penance for staying in the sun too damn long!*)

Unfortunately for the woman in the South of France, being in a fentanyl-induced coma rendered both of these finely tuned mechanisms irrelevant. Most of the time, though, our body has a very clear idea of how much light from the sun it wants to absorb, and it'll happily make us feel miserable to make sure we stay within those limits. But other than being cooked alive, what's it trying to avoid?

Before we get to that, let's start by looking at *how* sunlight cooked the unfortunate Brit.

The sun emits little packets of energy called photons. Each photon carries a very specific amount of energy, and a photon's energy determines *everything* about it, starting with whether you can see it or not. Our eyes are a pair of extremely sensitive photon detectors, and what you know as "light" is actually an unimaginably large number of photons on the last leg of their journey from the sun, bouncing off everything around you and crashing into photon-detecting proteins in your retina.* This collision generates an electrical signal that your brain interprets as whatever you happen to be looking at, e.g., two lions mating. Our eyes can detect photons within only a very narrow energy range: from 0.00000000000000000028 to 0.00000000000000000052

....

* Physicists would point out that light sometimes behaves as a continuous wave rather than individual balls of energy. This is true, but just to make life simple, I'm going to stick with the packet metaphor.

joules.* But the energy range of photons emitted from the sun is *much wider*: roughly 0.0000000000000000000000000020 to 0.0000000000000020 joules. Most of these, though, are helpfully absorbed by that thin layer of O_3 molecules you learned about in middle school called the ozone layer. What that means is that the woman in the South of France was bombarded with photons with energies from about 0.000000000000000000079 to about 0.00000000000000000068 joules per photon (10 times as much).

If you're going cross-eyed from all the zeros, here's a not-remotely-to-scale sketch of the photons—visible and invisible—that were crashing down on the woman in the South of France:

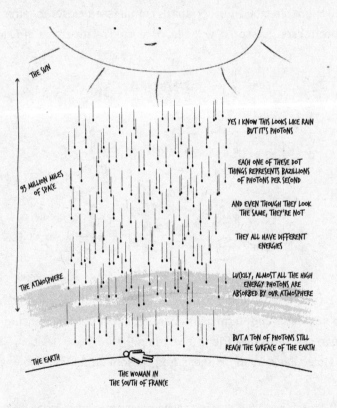

THE SUN

95 MILLION MILES OF SPACE

THE ATMOSPHERE

THE EARTH

THE WOMAN IN THE SOUTH OF FRANCE

YES I KNOW THIS LOOKS LIKE RAIN BUT IT'S PHOTONS

EACH ONE OF THESE DOT THINGS REPRESENTS BAZILLIONS OF PHOTONS PER SECOND

AND EVEN THOUGH THEY LOOK THE SAME, THEY'RE NOT

THEY ALL HAVE DIFFERENT ENERGIES

LUCKILY, ALMOST ALL THE HIGH ENERGY PHOTONS ARE ABSORBED BY OUR ATMOSPHERE

BUT A TON OF PHOTONS STILL REACH THE SURFACE OF THE EARTH

....

* Joules are units of energy—just like calories on a nutrition label. One calorie is a little less than 4,200 joules.

Each of those photons would have interacted with her body slightly differently. Let's start by looking at the subset of photons with slightly less energy than those you can detect with your eyes. These photons penetrate deeper into human skin than you might think: at least a millimeter but possibly more, depending on your skin tone and the exact energy of the photon.* That means they interacted with many, many, *many* of her cells and molecules inside those cells: DNA, proteins, sugars, fats, cholesterol, and water, to name a few. When these photons hit electrons in all those molecules, they made them *move* in all kinds of different ways. Whole molecules spun around and/or pairs of atoms within the molecules got stretched further apart, compressed closer together, or bent, rocked, scissored, wagged, or twisted relative to a third atom.

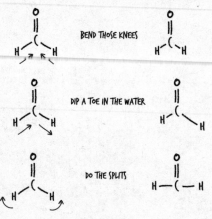

Basically, everything flailed about in a seemingly random, uncoordinated, and ungraceful manner, like a frat boy dancing his heart out at his best friend's wedding.

Molecular dancing is measured by something you know well:

....

* If it seems weird that light can get under your skin, try covering the front of a powerful flashlight with your hand, in the dark. You should still be able to see the light. That means photons from the flashlight's bulb are traveling all the way through parts of your hand and reaching your eyes.

temperature. The hotter something is, the more vigorously the molecules that make up that thing are dancing. So the molecules of water in, for example, a pot of boiling water are dancing much, much faster than the molecules of your hand cells. If you stick your hand into that pot of boiling water,[*] the water molecules will dance up against your skin molecules EXCEEDINGLY VIGOROUSLY, which will cause your skin cell molecules to dance much more vigorously than they were before.

Your nerve endings will sense this molecular dancing and send signals to your brain that get interpreted as *FUCKFUCK-HOTFUCKHOTHOTHOTJESUSCHRISTPULLHANDOUTNOW-FUCKFUCKSHITOUTHANDOUTPULLHANDOUTFUCKADUCK*, or something like that.

MY HAND

+

A MOLECULE IN MY HAND, NOT MOVING VERY MUCH

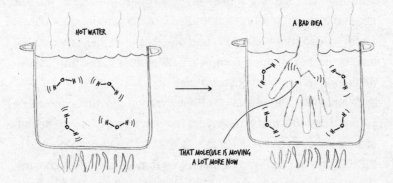

HOT WATER

A BAD IDEA

THAT MOLECULE IS MOVING A LOT MORE NOW

....

* Do not try this at home. Or anywhere else, for that matter.

Photons that make your skin molecules dance (like hot water does) are called "infrared light." Yup, that buzzword of the sun's warmth, thermal cameras, and super-fancy cooktops is just a label we created for photons with a very specific set of energies, and the word "warm" is just the label for what it feels like when they hit our skin.

Clearly, the total amount of energy you absorb from solar photons (per second) is much less than what you absorb by sticking your hand in a pot of boiling water (per second), which is why the photons cause a pleasant warming sensation and sticking your hand into boiling water causes horrible burning pain. But if you get hit by enough infrared photons, a wide array of bad things happen. Cells explode. Proteins coagulate and become useless. Eventually the water boils off and the solids combust, forming gases and carbon, which you know as the charring on your steak.

The woman who lay in a fentanyl-induced coma on the beach in southern France was hit with roughly 20,000,000 joules of energy—that's about what you would absorb if your body was cooked on a gas range for seventeen minutes.

At the same time as the woman in the South of France was being bombarded with infrared photons from the sun, she was also being bombarded with other photons, also from the sun, but with slightly higher energies than the ones you see: about 0.00000000000000000052 to 0.00000000000000000068 joules, called ultraviolet (UV) photons. These higher-energy photons caused something completely different—something *good*—to happen in her skin: photosynthesis.

I know, that sounds weird. Photosynthesis is supposed to be plants' domain—and it mostly is. But you can do it too. The outermost layer of your skin contains lots of 7-Dehydrocholesterol, or 7-DHC, one of cholesterol's much less famous chemical cousins. 7-DHC, when hit by these particular photons, changes into a molecule called previtamin D_3, setting off a chain of events that produces the active form of vitamin D. This is basically the same thing that happens in plants: the use of light to drive a chemical reaction. Unlike plants, we don't photosynthesize food, but we do photosynthesize an essential chemical that we need to survive.[*]

We kinda glossed over this when we talked about plant photosynthesis, but it's actually pretty weird: light can drive a chemical reaction; in other words, light can change one molecule into another. It can change the very nature of a substance. How does it do that? One way is via heat. Infrared light can cook flesh: that definitely qualifies as changing the nature of a substance. Ultraviolet light, not to be outdone, can excite a molecule's electrons, which in turn can break chemical bonds and form different ones instead, thus changing one substance into another.

When it comes to changing previtamin D_3 into vitamin D, that's a good thing.

But it can also be bad.

DNA is the best-marketed molecule in the world. If *DNA* were the title of a movie, the subtitles might read:

YOUR GENETIC CODE

or

THE BLUEPRINT FOR YOUR LIFE

or

THERE IS NO SUCH THING AS FREE WILL.

....
* Incidentally, nowadays we add vitamin D to all kinds of stuff—mostly milk—to avoid vitamin D deficiency diseases like rickets or osteomalacia.

As with anything marketed to within an inch of its life, it's worth looking past the advertising copy and seeing what DNA actually looks like.

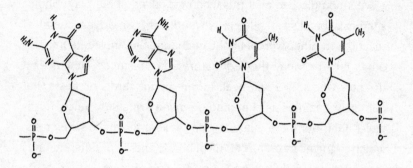

Remember, the lines represent electrons shared between two atoms: chemical bonds. So what you're seeing here is how every single atom that makes up DNA is bonded (or not) to its neighbors. I know this looks like an ungodly mess, but we can simplify it. See that repeating pattern across the bottom? That's what you've probably heard of as the "backbone." It's a series of sugar and phosphate molecules:

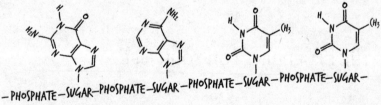

Okay, that's better. But we can go further. Those things that are attached to the sugars are called bases, and there are four different types of them in DNA: adenine, thymine, guanine, and cytosine. You probably know them as the (fanfare and trumpets!) Letters of the Genetic Code, so we might as well convert them to their shorthand labels: A, T, G, and C. Let's also convert sugar and phosphate to S and P:

```
   G      A      T      T      A      C      C      A
   |      |      |      |      |      |      |      |
—S—P—S—P—S—P—S—P—S—P—S—P—S—P—S—
```

Biologists condense this even further, to a single stream of letters, which, in this example, would be GATTACCA. When people say that your genetic code is "three billion letters long," this is what they mean: there are three billion of these A's, T's, G's, and C's that instruct your cells on how to build themselves and conduct every aspect of their lives.[*]

Notice how all the bonds are laid out. Phosphates are bonded only to sugars. Letters are bonded only to sugars. Letters are not bonded to adjacent letters. Your DNA is an exquisitely intricate coding system, but the code itself is dependent on how DNA's different components are bonded (or not bonded) to each other. And since DNA's bonds are electrons, what that means is that DNA can't do its job unless its electrons are in the right places.

When incredible numbers of ultraviolet photons descended from the heavens and landed on the woman in the South of France, the ones that crashed into DNA's electrons excited some of them. (Remember, exciting electrons can make molecules more likely to chemically react, like the previtamin D_3.) Thankfully, most of the time DNA is hit by photons, the excited electrons just come back down to their non-excited state, and your DNA

....

* Technically, this is true for each strand of DNA. Every cell has two strands, so your genome is really made up of six billion letters.

remains chemically the same as it was before. I know, kind of a letdown, right? But this is really good for your health. Remember that even if you're only under the sun for a few minutes, you're getting bombarded with bazillions of UV photons per second; if most or even some of them had any permanent effect on your DNA, you'd be in serious trouble.

But very, very rarely, a photon changes the way bonds are arranged in your DNA. It might end up looking like this:

See those two C's? They're now fused together, like messed-up spinal vertebrae. You might think that's not a big deal; after all, there are six billion other letters in a cell's genome, so how much harm could this possibly do?

It can kill.

Most of the time, your cell detects this fused-C problem and repairs it.* You go on with your life as if nothing happened. Remember from the last chapter that this is actually the best-case scenario. Occasionally, the repair effort fails catastrophically, in which case the cell goes, *I'M OUT,* and kills itself. This is the medium-case scenario. The worst-case scenario—which is relatively rare—is that at some point during the repair or even cell division process, the cell copies its DNA wrong. So a sequence like "GATTACCA" might become "GATTA**TT**A." In other words, the

....

* Proteins unzip the two strands of DNA, cut away the sugar-phosphate backbone of the damaged strand before and after the fused C, read the other strand to figure out what the correct sequence should be, and rebuild the cut strand. Incidentally, this is why having *two* strands of complementary DNA is useful. If one breaks, you haven't lost the information.

worst-case scenario is that a *mutation* might creep into the DNA sequence.

Once a mutation gets into your DNA, it's pretty much locked in. Your body can detect and fix a fused-C. But once that fused-C becomes, say, a TT, your body can no longer detect or fix it, because that mutation is *chemically* healthy DNA (even though it's *informationally* diseased). I know: that seems strange. But consider this: in 2012, the *Centralia* (IL) *Morning Sentinel* reported that Eric Lyday, the bandmate of a local musician, was "on drugs." What the paper actually meant was that Eric was "on drums." That one-letter typo entirely changed the meaning of the sentence . . . without breaking any grammatical or spelling rules. DNA mutations do the same thing: they change the meaning of the DNA without breaking any of its chemical rules.

Did the woman in the South of France get a mutation from lying under the sun for six hours? Probably. Hmm . . . so it's kind of weird that the medical literature didn't mention *HER NEW SU-PERPOWERS.*

Unfortunately, no matter how many genetic mutations you get from lying in the sun, you won't suddenly start glowing bright green or get superpowers. In fact, most mutations don't do anything at all. Only about 1 percent of your genome actually codes for a protein, so if a mutation happens elsewhere, it's probably harmless. Or if you get a mutation in, for example, a cell in your intestinal lining, that's probably not a big deal, either, because you'll poop that cell out pretty quick. But we do know that the more mutations a cell gets, the more likely it is to become cancerous, so anything that increases your natural mutation rate—including too much sun—is not great.

Let's pause here and take stock. Everything we talked about so far is part of a Bridge of Truth that connects the sun to skin cancer. And just like the Bridge of Truth that connects smoking to lung cancer, this bridge also has different kinds of bricks. All the bricks we've talked about so far are molecular: they show *how* UV light interacts with the molecules in your skin and leads to skin cancer. But just like with smoking, the molecular bricks came later. The bridge was built from non-molecular bricks first. Let's look at a few.

First: where you work. All the way back in the late 1800s and early 1900s, scientists noticed that farmers, sailors, and others who spent their lives mostly outdoors had much higher rates of cancer than similar people who lived and worked in cities. One doctor who practiced in a mining town reported that in his twenty-five years, he'd seen only two cases of skin cancer among miners.* Today, we can quantify the difference in risk: if you work outside, your risk of skin cancer is roughly 300 percent as high as an indoor worker.

Second: clothing. Unless you're a rabid exhibitionist, you probably wear clothes. Clothes absorb some ultraviolet light, so you'd expect that skin cancers would be less likely to develop on parts of the body that are typically covered up . . . and you'd be right. Skin cancers are much less common on, say, your feet, thighs, and butt than on your scalp, ears, or nose.

Third: skin color. Skin cancer is *much* more common in white people. It's hard to calculate exact relative risks, but estimates put the risk of skin cancer in white skin between 1,600 percent and 6,300 percent as high as in black skin. Why? Melanin. Melanin is a molecule your body produces and distributes throughout your skin, and it absorbs ultraviolet photons LIKE A BOSS, thus preventing them from potentially damaging your DNA. So the more melanin in your skin, the less DNA damage—and skin cancer— you'd expect. (Melanin *also* absorbs visible light, which means the more melanin in your skin, the darker it is. But dark skin does

....

* Unfortunately, he didn't make the connection to the sun. He thought the miners were spared because they drank a lot of tea.

not make you immune to skin cancer—and unfortunately, it can make cancer harder to detect. So even if you have darker skin, you should still be vigilant.)

And last, the crème de la pièce de résistance. Remember in chapter one, we discussed a randomized controlled trial of processed food? We were going to split up thousands of people into two groups, send each group to their own deserted island, feed one group ultra-processed food and the other unprocessed food, and follow them for fifty years. Turns out, Great Britain set up a remarkably similar experiment—to test whether the sun causes skin cancer—except their experiment was accidental. Chances are you already know about this experiment. It's called "Australia." Between about 1788 and 1868, the British sent more than 150,000 convicts to Australia. In other words, they took a bunch of genetically similar people (their population), split them up into two groups (convicts and . . . not convicts), and put each group on its own island (Great Britain and Australia). Because Australia is much closer to the equator than Great Britain, and because the Australian sky is not a wet gray blanket woven from mist and despair, Australia gets hit by *way* more ultraviolet photons than Great Britain. And because the Australians were—and still are—mostly white, they didn't have much melanin to protect themselves from all those photons. So you'd expect to see a lot more skin cancer in Australia than England. And you do. To this day, an Australian's risk of developing at least one skin cancer in their lifetime is about 660 percent as high as a Brit's.[*]

We could spend the rest of this chapter looking at bricks in the sun/skin cancer Bridge of Truth (noting, for example, that for many of them, the story is complicated by the different *types* of skin cancer), but let's move on.

So, quick recap on photons from the sun: plants use them to

....

[*] Assuming, of course, that each of them lives in their respective country.

make food; we use them to make vitamin D. But too many photons can burn us, or even cause skin cancer.

Now we get to the part of the book where I tell you that humanity's multi-millennial experience with processing nature—from detoxifying plants to make them into food, to preserving that food, to turning aphid shit into candy—prepared us to synthesize our knowledge of the sun and skin cancer and design the perfect consumer product to solve this human health need: sunscreen.

Er . . . not quite . . .

Almost every bottle of sunscreen at the drugstore says it'll reduce your risk of skin cancer, but that's not why sunscreen was invented. In fact, sunscreen is much, much older than our understanding of skin cancer. People were processing stuff from nature to make sunscreen millennia ago. For example, ancient Greeks and Egyptians smeared all kinds of things—like oil, myrrh, and rice bran—all over themselves to try and prevent tanning.

But the roots of modern sunscreens can be traced to a single product: Ambre Solaire, created by Eugène Schueller in 1935. Back then, the link between the sun and skin cancer was not well understood. In fact, Ambre Solaire was invented nine years before anyone realized that DNA carried our genetic information, eighteen years before we knew the structure of DNA, and more than forty years before we knew that cancer could be caused by DNA mutations. That's because Ambre Solaire was invented to try and prevent sunburn, not skin cancer. In 2012, the FDA's sunscreen labeling rules officially went into effect, specifically allowing manufacturers to claim that a sunscreen "decreases the risk of skin cancer." To figure out why the FDA allows manufacturers to make this claim, let's look at two of the most common active ingredients in sunscreens sold in the United States: zinc oxide and oxybenzone (also known as benzophenone-3).

You might have read that zinc oxide is a type of "physical" sunscreen and oxybenzone is a type of "chemical" sunscreen, and that the former reflects photons like a shield and the latter absorbs them like Whitney Houston's bodyguard absorbs bullets in her Oscar-nominated hit *The Bodyguard.*

That's wronger than an Oreo in orange juice. What they actually do is a lot weirder. Let's look at oxybenzone:

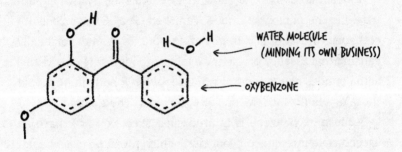

WATER MOLECULE (MINDING ITS OWN BUSINESS)

OXYBENZONE

To give you a sense of size, there would be roughly 700,000,000,000,000,000,000 molecules of oxybenzone in a typical quarter-size splooge of sunscreen, and if you apply the recommended dose to your skin, you'd be spreading about 8,400,000,000,000,000,000,000 molecules of oxybenzone over every square inch of your exposed body.

When an ultraviolet photon from the sun hits a molecule of oxybenzone on your skin, it sets off a somewhat complicated chain of events. First, the photon crashes into an oxybenzone molecule, putting it in an excited state, which just means it has more energy than it did before. The molecule looks the same:

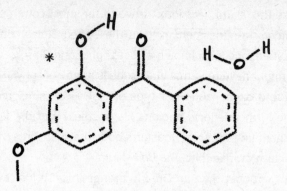

We just add a little * to show that excited state. But what happened to the photon? It's gone. Vanished. *Poof.* Oxybenzone absorbed it, preventing it from hitting your DNA and potentially causing that fused-C problem we talked about before. So far, this actually does sound kind of similar to what a bodyguard would do: take a bullet for someone else. But wait. There's more.

Because oxybenzone is in an excited state, you now have an excited-state molecule on your skin, which might be just as damaging as having a high-energy photon hit your skin. But oxybenzone can get rid of that extra energy through the power of *DANCE*!

First, the molecule does this:

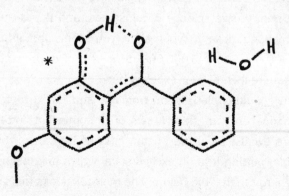

(the "wine glass juggle").

Then this:

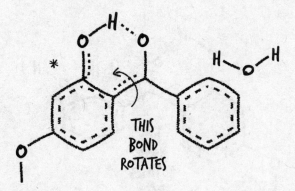

(the iconic "*nae nae* stanky leg").

And this:

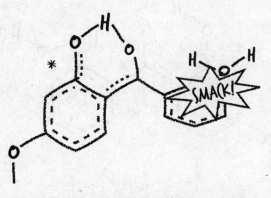

(the "Shrek hip bump"). Notice that the molecule next to oxybenzone gets smacked in this step.

And then this:

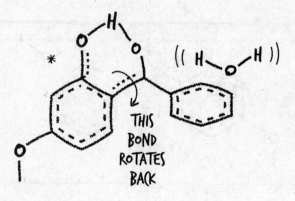

THIS
BOND
ROTATES
BACK

And we're almost back to where we started:

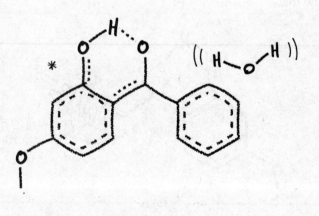

Much like sloppy human wedding dancing, oxybenzone bumps into other nearby molecules, transferring some of its molecular motion to them, thus heating up its immediate environment.

Notice that oxybenzone has managed to dance itself back to the way it was at the very beginning, before it was hit by a photon. So this series of sloppy dance moves that generate heat is actually a *cycle*: an ultraviolet photon goes in; heat comes out.[*]

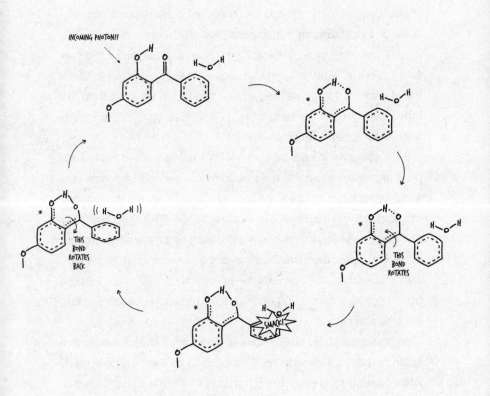

....

[*] But wait: if sunscreen converts light energy to heat energy, does wearing sunscreen make you hotter when you're in the sun? Probably. But remember, your body is also being hit by an unfathomable number of infrared photons, which directly heat your skin. There's so much direct heat from infrared photons that you wouldn't feel the teensy bit of extra heat from the ultraviolet photons heating up your sunscreen.

Zinc oxide and titanium dioxide (the so-called physical sunscreens) also cyclically absorb photons and convert them into heat energy, though the exact mechanism is different. Health blogs, news articles, and even dermatologists say they "reflect" or "scatter" UV light. In fact, some sources suggest they reflect or scatter only as little as 5 percent of UV light and absorb the rest. My suspicion is that the confusion arose because some formulations of zinc/titanium sunscreens look like white cream cheese spread out on your skin. People just assumed that, since the sunscreens were scattering visible light—making you look like a bagel waiting for its salmon—they must also be scattering UV light. But whether something reflects visible light can be unrelated to whether it reflects UV light.

Back to oxybenzone. Its convert-UV-photon-to-heat cycle happens *fast*: it takes roughly ten-trillionths of a second for a molecule of oxybenzone to get back to the way it was.[*] This means that one molecule of oxybenzone can absorb about 90,000,000,000 UV photons per second. If you apply the FDA-recommended amount of SPF 30 sunscreen, what you're doing is enhancing your skin's ability to harmlessly dissipate the energy from well over 700,000, 000,000,000,000,000,000,000,000 ultraviolet photons crashing into you per second.

So, to summarize: our species has engineered a creamy white splooge that you spread over your body to convert the potentially DNA-damaging energy from hundreds of million septillions of ultraviolet photons per second into mostly harmless heat.

In case you're wondering, sunscreen would not have helped the woman in the South of France. The photons responsible for

....

[*] You might be asking: How the hell do we know all this? The answer is pump-probe spectroscopy, which can "see" things that happen on picosecond time scales. (One picosecond is the time it takes for a photon of light to travel one-third of a millimeter.)

heating her skin to the point of roasting it were infrared, not ultraviolet, and thus wouldn't have been absorbed by the sunscreen. But even if they had been, the woman was in the sun for six hours; any sunscreen would have been overwhelmed by the sheer number of photons hitting her.

On one level, modern sunscreen isn't so far from smearing yourself with clays, minerals, or a mixture of sand and oil like the ancient Egyptians or Greeks did. But on another level, modern sunscreens are some mind-bending magico-chemical spellwork.

Our species should be patting ourselves on the back right now.

But does our little magic trick actually work?

———

That's not just a philosophical question. It's a practical one. Let's say you're at the drugstore buying a bottle of sunscreen because your dermatologist threatened to go on a hunger strike if you didn't. Which one do you pick? Nobody would blame you for spending hours in the sunscreen aisle, completely confused. Bemused. Befuddled. Overwhelmed.

It's not you. Sunscreens bear the most incomprehensible labels you're likely to be confronted with. A representative example is on page 140.[*]

It doesn't seem like it, but the label actually contains many of the clues we'll need to figure out the practical (and philosophical) questions of whether a sunscreen works.

Let's start with SPF. Both Merriam-Webster.com and the *Oxford English Dictionary* define "SPF" as "sun protection factor." Both of these storied repositories of our hallowed English language have

....
[*] This sunscreen is fictitious, and any resemblance to a real sunscreen is entirely coincidental.

it wronger than peanut butter on pepperoni. "SPF" should really stand for "sun*burn* protection factor." (Remember, Ambre Solaire was invented so that pasty white Europeans could get a sun*tan* without risking a sun*burn*.)

SPF is kind of tough to wrap your head around. The first thing to know is that it's not spit out by an algorithm; it's a quantity that some unfortunate person in a nondescript medical office building somewhere actually measures. The procedure, which is mandated by federal law, goes roughly like this:

1. Find a white person (not off-white or cream; they have to be printer-paper-white).*

....

* The FDA requires that sunscreen be tested on a person who "always burns easily" or "burns moderately" during the first thirty to forty-five minutes of sun exposure after a winter season of no sun exposure. A person who "always tans well," "tans profusely," or is "deeply pigmented"—in other words, brown or black people—is ineligible for sunscreen testing. Europe has a similar regulation. That doesn't mean, of course, that people with darker skin don't get sunburned or shouldn't wear sunscreen. There's a wide range of sensitivity to sunburn even among people of similar skin tones. Light skin doesn't necessarily doom you, and dark skin doesn't necessarily protect you.

2. Cut out a stencil with two rows of rectangular boxes and put it over their lower back.

3. Smear a very specific amount (2.0 milligrams per square centimeter) of sunscreen through the bottom row onto their back and wait for it to dry.

4. Using a lamp that is designed to put out only ultraviolet light, give this white person increasing doses of ultraviolet light (as you go left to right on the stencil).

5. Wait a day, and then see how much ultraviolet light was needed to just barely give them a sunburn on the top row (no sunscreen) vs. the bottom row (with sunscreen).

6. Then calculate the SPF like this:

$$SPF = \frac{\text{AMOUNT OF UV LIGHT REQUIRED TO JUST BARELY GIVE THE WHITE PERSON A SUNBURN } \textit{WITH} \text{ SUNSCREEN}}{\text{AMOUNT OF UV LIGHT REQUIRED TO JUST BARELY GIVE THE WHITE PERSON A SUNBURN } \textit{WITHOUT} \text{ SUNSCREEN}}$$

7. Repeat with a bunch more white people and take the average of the SPFs you found.

So if you're in the drugstore and holding two bottles of sunscreen in your hands, one SPF 25 and one SPF 50, you know that both sunscreens have been tested in a lab somewhere, *by* humans and *on* humans, and that the SPF 50 lets in roughly half the amount of sunburn-causing ultraviolet energy as the SPF 25 sunscreen. This is true of every legitimate sunscreen product in every major market in the world. So sunscreen really *does* work, in the sense that it unequivocally reduces your risk of sunburn.

When it comes to actually interpreting what the SPF *means*, we sometimes have trouble. Have you ever heard something like: *If it takes 20 minutes for your unprotected skin to start turning red, using an SPF 15 sunscreen theoretically prevents reddening 15 times longer—about five hours.* This is *sorta technically* true, but unfortunately, it leads to people doing math like this:

$$\text{THE TIME IT NORMALLY TAKES ME TO BURN} \times \text{SPF} = \text{SUNBURN-PROOF TIME IN THE SUN! YAY!}$$

Suppose you think it takes you twenty minutes to burn without sunscreen. If you slather on SPF 100, you might think you can gallivant in the sun for *thirty-three hours* and not get burned. That's some hot nonsense. Here's why: First, you have no idea what the "time it normally takes me to burn" is. Second, that number is not fixed. It changes dramatically based on the time of day, time of year, where you are on Earth, what's underneath you (sand? snow?), and what's above you (clear sky? clouds?). And third, you almost never get the full protection of the SPF listed on the label. Why? Many reasons, the simplest of which is: You almost never apply as much sunscreen as they use in the official test, 2 milligrams per square centimeter of skin.

That's *a lot* of sunscreen. I tried putting that much on one summer and I felt like I had walked through an I Can't Believe It's Not Butter! carwash. For this reason, most people apply half this amount or less. And this leads to another misconception: that people put on "too little" sunscreen. This is . . . meaningless.

Nobody tells you how much butter to put on yourself; you just go for however much feels right. Same with sunscreen. Just be aware that "what feels right" is probably about half what the FDA mandates. That's actually one reason the bottle says to reapply: because it knows you didn't put on "enough" the first time around.

Another very popular—and also wrong—interpretation of the SPF goes something like this: *Once you get above SPF* [insert random number between 10 and 30 here], *the number doesn't really make much difference.* This myth is in the *New York Times* and *Consumer Reports*, on Gizmodo and the *Encyclopædia Britannica*'s website, and in peer-reviewed scientific articles written by practicing dermatologists. And everybody's reasoning is very similar. It's based on a table showing what percent of sunburn-causing UV light is absorbed by sunscreens of different SPFs:

SPF	PERCENT OF SUNBURN-CAUSING UV LIGHT ABSORBED BY THE SUNSCREEN
1	0%
15	93.3%
30	96.6%
50	98.0%
100	99.0%

Well-meaning people look at the table above and they write sentences like these:

> *An SPF of 15 blocks about 93 percent of the UVB radiation, while an SPF of 30 blocks out 97 percent of the UV radiation. This is only a 4 percent difference . . .*

This is wronger than meat loaf at a clambake. To see why, let me try to sell you a couple "bullet-resistant vests." Vest A stops 93 percent of bullets. Vest B stops 97 percent of bullets. It *seems* like there's only a 4 percent difference between the two vests, but consider this: If someone shot one hundred bullets at you and you were wearing vest B, you'd be hit by three bullets. In vest A, you'd be hit by *seven*—more than *double* vest B. Ditto with photons: the number of photons *blocked* by the sunscreen is totally irrelevant. The number that matters is how many *get through.*

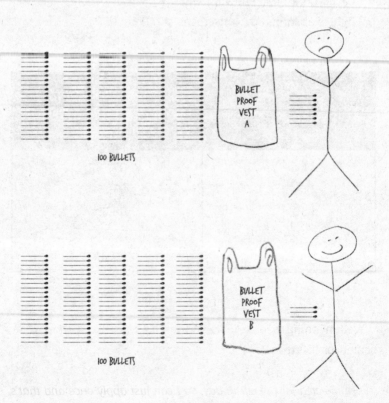

100 BULLETS

100 BULLETS

With that in mind, let's add a column to the table above:

SPF	PERCENTAGE OF SUNBURN-CAUSING UV LIGHT ABSORBED BY THE SUNSCREEN	PERCENTAGE OF SUNBURN-CAUSING UV LIGHT THAT GETS THROUGH THE SUNSCREEN
1	0%	100%
15	93.3%	6.7%
30	96.6%	3.4%
50	98.0%	2.0%
100	99.0%	1.0%

There. Now we have a much better sense of how two different SPFs relate to one another: you can see that SPF 100 absorbs *twice* as many sunburn-causing photons as SPF 50, and SPF 30 absorbs twice as many as SPF 15 (assuming you put the same amount of sunscreen on, of course).

So should you go for the highest SPF available? In the late 2000s, sunscreen manufacturers certainly thought so: they were constantly trying to outdo each other by making ever-higher-SPF sunscreens. I tend to go for the highest SPF I can find, but this is definitely not a one-size-fits-all approach. There are legitimate reasons you might not want to use ultra-high SPF sunscreens. Using a lower SPF sunscreen might be a good way to psychologically trick yourself into reapplying.

Wait. What?

The logic goes like this: if you apply SPF ELEVENTY BILLION sunscreen, you might think, *Oh, this is plenty to keep me 100 percent protected for the entire day, so I can just apply once and that's* *it.* Unfortunately, that's not true. Any sunscreen—no matter what SPF—will eventually get washed away by all your beachy activities, or toweled off, or diluted by your sweat. So if you're going to spend all day in the sun,[*] you need to reapply. If, however, you're using only SPF 30, you wouldn't feel so protected, so you'd consistently reapply it throughout the day.

By the way, you may have noticed that sunscreen labels tell you to "apply liberally 15 minutes before sun exposure."

Why?

Because sunscreen is not moisturizer. You don't want to rub it *under* the top layer of your skin; you want it to form a protective barrier on *top* of your skin. So, contrary to basically everything you've been taught your entire life, the right way to put on sunscreen is to spread it very lightly over the surface of your skin, then let it dry. As it dries, it binds to the top layer of your skin. That's what the fifteen-minute waiting period is for. If you put on sunscreen and then immediately put on your clothes, you might unintentionally wipe it off before it has a chance to bind to the top layer of your skin.

Does sunscreen work?

It absolutely *does* reduce your risk of sunburn. That's crystal clear, because every commercial sunscreen is smeared on a person and the SPF is calculated by actually observing how much more ultraviolet light it takes to give that person a sunburn while they're wearing sunscreen.

But the picture is a little muddier when it comes to skin cancer.

There are two basic types of skin cancer: melanoma and non-melanoma. Almost all skin cancers are non-melanoma, which can

....

* Which you shouldn't. More on that soon.

be further divided into either squamous cell carcinoma (SCC) or basal cell carcinoma (BCC). If you *had* to get cancer but could choose the type, you'd pick BCC: it's very slow growing and rarely metastasizes. On the other hand, melanoma is much more serious. It accounts for a minority of skin cancer cases but causes most of the deaths.

We absolutely know that the sun *causes* skin cancer. The question is whether using sunscreen *protects* against it. Intuitively, it seems like it would: we know it absorbs sunburn-causing ultraviolet photons. But as cancer researcher John DiGiovanna says, "Sunscreen isn't a suit of armor. It can be overcome by too much sun." Unless you're submerged in a pool of sunscreen, some solar photons will absolutely get through to your skin; that's one reason the FDA doesn't allow manufacturers to use the word "sunblock." But there's also this:

1. Photons have different energies
2. different-energy photons can do different things to your skin
3. and different sunscreens may absorb photons of differing energy differently.

That's a mouthful. Let's break it down.

In Copenhagen in 1932, at the Second International Congress of Light—which sounds like some sort of Illuminati gathering—a bunch of physicists got drunk and created arbitrary divisions within ultraviolet light. You've almost certainly seen these arbitrary divisions before: they're called UVA and UVB, and maybe your dermatologist explained them to you roughly like this:

UVB CAUSES SUNBURN (AND SOME CANCERS).

UVA CAUSES WRINKLING (AND SOME CANCERS).

This is not exactly true, but a perfectly fine simplification for our purposes. Early sunscreens absorbed UVB photons very well and absorbed UVA photons . . . not so well. You might call these sunscreens "narrow spectrum." And narrow spectrum works great for shielding against sunburn-causing UVB photons, but to protect against a fuller range of the sun's photonic assault you also need to absorb UVA photons. Hence the BROAD SPECTRUM on the label.

The FDA allows any sunscreen that has an SPF of 15 or higher that also passes its broad-spectrum test to say that it "decreases the risk of skin cancer . . . caused by the sun." What's the evidence for that claim?

Um . . .

Well . . .

It's kind of embarrassing to admit this, but so far there has been only *one* randomized controlled trial that tested whether sunscreen could reduce the risk of skin cancer, and that trial was mostly focused on non-melanoma skin cancers. It found that sunscreen didn't change the number of *people* who got squamous or basal cell carcinomas, but it did reduce the *number* of squamous cell carcinoma tumors diagnosed per person. This is not exactly the type of ironclad evidence you'd hope for, though I will point out two factors in this trial's defense. First, it was conducted in the 1990s, which means it used pretty old sunscreen technology. If we redid the trial with modern sunscreens, we would expect a more dramatic result. Second, the control group in the trial was not prevented from using sunscreen; that would have been unethical. They were allowed to use sunscreen, but they used less than the full-on-sunscreen group. If people in the control group had been prevented from using sunscreen, we would also expect a more dramatic result.[*]

....

[*] In this particular case, a more dramatic result would have been people in the control group getting a lot more cancer. This would have been bad experimental design on many fronts: First and most obviously, preventing

What about melanoma? Again, the evidence here is . . . less than ideal. The only randomized controlled trial on melanoma in adults was actually a continuation of the trial we just talked about. Both this trial and a couple of cohort studies suggest that sunscreen does have a protective effect.

Data on melanoma rates reveals a bit of a paradox: even though lots of white people throughout the world use sunscreen, melanoma rates have not gone down or even stayed flat. In fact, over the past thirty years, they've *tripled*. If sunscreen protects against skin cancer, why are melanoma rates rising?

One explanation could be that people enjoy tanning and burning the living crap out of themselves more than they used to, so even though they use sunscreen, they also expose themselves to *way* more sun than they used to. Under this hypothesis, melanoma rates would be even higher if people didn't use sunscreen.

But there's another hypothesis. It was advanced by a Belgian epidemiologist named Philippe Autier, and although it's supported by two (small) randomized controlled trials, it remains controversial. Autier believes that sunscreen use among white people who like to sunbathe actually *increases* total UV exposure, which could lead to melanoma. His thinking goes like this: White people like to intentionally expose themselves to the sun to get a tan, but they don't like to burn. So they buy ultra-high-SPF sunscreen, which effectively absorbs most of the photons that cause sunburn. But because they're *not* getting sunburned, these white people stay out in the sun much longer than their bodies would otherwise let them.

Basically, Autier believes that sunscreen lets you circumvent your biochemical "GTFO of the sun!" alert, thus allowing you to

....

people from using something that might reduce their risk of cancer is unethical. Second, it would have made the trial results look better, but more people might have ended up with cancer than if there had been no trial. And third, it wouldn't have changed the sunscreen's actual effectiveness; all it would have done was made it look better by comparison.

overdose on sun exposure. He goes so far as to say that the recommendation to reapply sunscreen—which is required by law in the United States—"probably represents a form of abuse."

That's pretty wild.

But where does it leave us?

As sunscreen expert Brian Diffey told me, it leaves us with "a dilemma." On the one hand, the evidence that sunscreen protects against skin cancer is not as robust as evidence would be for, say, a new cancer drug. But on the other hand, we know that photons from the sun cause skin cancer, and we also know that our bodies do not respond well to too much sun. So what's my take-home message here? Basically this: it's better to avoid getting hit with ultraviolet photons. I don't tan myself for fun, whether in the sun or in a tanning bed, and when I'm outside, I try to stay in the shade. Does this mean I avoid the sun like a vampire? Absolutely not. We all need a baseline level of ultraviolet light to make vitamin D (assuming you're not getting it from your diet). Plus, sometimes the sun just feels so damn good.

If, for some reason, you have to be in the sun for a long time, should you wear sunscreen?

Here, I'd say . . . yeah, sure. Sunscreen will reduce the number of ultraviolet photons interacting with the molecules in your skin, and that might reduce your risk of skin cancer. So I say go for it. But I also think it's a good idea to wear a hat. And clothes. And a burkini wouldn't hurt, either.

Final question: Should you make sunscreen part of your everyday routine?

This one's a little more complicated.

Are the chemicals in sunscreen bad for you?

Unless you're allergic to any of them, the answer is: not immediately. But what about over long periods of time—for example, if you were to religiously put on sunscreen every day for thirty years? If you were to spend a few hours googling sunscreen safety, you'd find enough reading material to last you hundreds of poops. Let's dip our toes into some of this material. First let's review studies on some of the most common active ingredients in sunscreen: oxybenzone, octinoxate, octocrylene, zinc oxide, and titanium dioxide.

Oxybenzone, also known as benzophenone-3, can soak through your skin and get into your pee, breast milk, and bloodstream, and once there, mimic your hormones. Animals exposed to oxybenzone and octinoxate, another ultraviolet absorber, were found to have lower sperm counts and higher sperm abnormalities. Female mice exposed to oxybenzone had messed up menstrual cycles. A recent study found that adolescent boys with higher oxybenzone levels had significantly *lower* testosterone levels.[*] Another study found a link between higher concentrations of benzophenones and poorer reproductive success in men trying to conceive via fertility clinics. And as if that weren't enough, oxybenzone has been shown to damage the DNA of coral larvae and cause reef bleaching and death. Hawaii banned both oxybenzone and octinoxate in 2019, and REI pledged to stop selling sunscreens with oxybenzone in 2020.

But don't believe for a second that if you avoid oxybenzone and octinoxate, you'd be out of the woods. According to the Internet, eight of thirteen UV filters approved for use in the U.S. have been shown to affect calcium signaling in male sperm cells. Homosalate, yet another ultraviolet absorber, can enhance the absorption of herbicides through your skin and into your blood. Octyl

....

[*] Though that could just be because higher-testosterone jocks are more likely to think, *Sunscreen is for wusses!* while stealing lower-testosterone, sunscreen-slathered nerds' lunch money.

methoxycinnamate has been shown to decrease motor activity in female rat babies, and 4-methylbenzylidene camphor has been shown to reduce female rats' sex drive and impair early muscular and brain development. Octocrylene messes with the expression of genes related to development and metabolism in the brains of zebrafish.

And lest you think that using metal oxide sunscreens instead of oxybenzone, avobenzone, etc., sunscreens will save you: both zinc oxide and titanium dioxide nanoparticles have been shown to affect spatial cognition in rats, impair learning and memory and increase reactive oxygen species in mice, decrease acetylcholine esterase activity in fish, decrease brain weight in honeybees, reduce cell viability in human brain cells, increase oxidative damage to the hippocampi of male rats, and decrease hatch time and increase malformation rate in zebrafish.

Parabens, another set of common sunscreen ingredients, have also been shown (according to the Internet) to disrupt the endocrine system, which could increase your risk of reproductive toxicity. Oxybenzone, benzophenone-4, avobenzone, octyl methoxycinnamate, octisalate, and octocrylene have all been associated with contact allergies, and methylisothiazolinone has received the dubious honor of being named "Allergen of the Year" by the American Contact Dermatitis Society in 2013. And as Hillary Peterson, a health blogger, points out, there are "5,000 different chemicals (including hormone-like and hormone-disrupting phthalates and synthetic musks) that can hide under the name 'fragrance,'" which can, when exposed to UV light, cause cell damage or death.

Now let's consider retinyl palmitate and its chemical cousins retinyl acetate, retinyl linoleate, and retinol. The "retin-" part of all those names refers to vitamin A, which we all need to eat a certain amount of to stay alive, and which cosmetics manufacturers have been putting into sunscreens (and also anti-aging creams, lotions, and foundations) for years (because vitamin A is a pretty decent antioxidant and studies have shown that it prevents wrinkles). Unfortunately, too much vitamin A can cause liver damage, brittle nails, and hair loss, and can contribute to osteoporosis in older adults and skeletal birth defects in developing fetuses. But the pièce de résistance of vitamin A's toxicity[*] is the fact that, when spread over the skin and hit with UV photons, it significantly increased the number of skin tumors and lesions in mice. Of the 500 sunscreens tested by the nonprofit Environmental Working Group back in 2010, over 40 percent contained vitamin A. That number is lower as of 2019 (about 13 percent), but still.

In early 2019, the FDA addressed some of these concerns. They published a proposed rule change indicating that twelve currently marketed sunscreen ingredients (including oxybenzone and avobenzone) might not be GRASE.

What does "might not be GRASE" mean?

GRASE is government acronym soup for "generally recognized as safe and effective," but the more important part of the sentence is actually "might not be." This is the FDA acknowledging that it doesn't have enough data to decide whether these twelve ingredients are safe and effective. You might reasonably ask: *What the eff? Shouldn't the FDA have figured this out years ago, when these ingredients were first put into sunscreens?!*

I will readily admit, this story line does not make the FDA look good. But in their defense, people have dramatically changed the way they use sunscreen. In the "old" days, you'd put it on only when you were planning to spend all day in the sun at the beach—maybe for a couple weeks every year. Now companies regularly put sunscreen into products you're supposed to wear

....

[*] One of the bits of knowledge I remember best from my four years at MIT is: If you ever kill a polar bear, don't eat its liver. This is because polar bear liver contains insanely high levels of vitamin A—enough to kill you if you eat the whole liver at once.

every day, and some dermatologists tell you to slather it on daily, forever. This means you're getting a much higher dose of all the chemicals in sunscreen than you were before. So the FDA was like, *Whoa, Nellie! We don't know what daily exposure to most of these chemicals over a period of years does.* In two cases, though, the FDA determined that there *was* enough information to declare a chemical GRASE: both zinc oxide and titanium dioxide can proudly wear that badge.

Woof. One innocent-looking bottle on the shelf, so many things to worry about! Here's a handy-dandy Diagram of Worry so you can keep track of everything you have to juggle in your brain when deciding whether to buy a bottle of sunscreen at your drugstore:

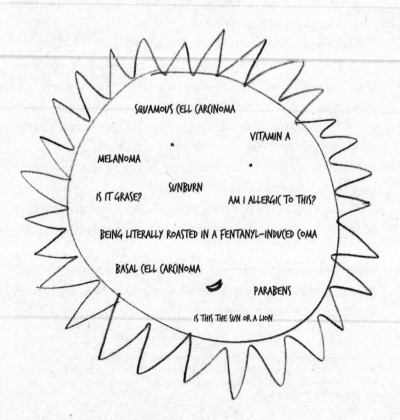

SQUAMOUS CELL CARCINOMA

VITAMIN A

MELANOMA

IS IT GRASE?   SUNBURN

AM I ALLERGIC TO THIS?

BEING LITERALLY ROASTED IN A FENTANYL-INDUCED COMA

BASAL CELL CARCINOMA

PARABENS

IS THIS THE SUN OR A LION

All this worry over one little product.

Whether or not to use sunscreen is, of course, just *one* of the 1,572 daily choices about what to put in your mouth, breathe into your lungs, or smear on your body.

We've stayed away from food for long enough. Let's ease back into food, in the same way that you might start a lazy Sunday: with a hot, steaming cup of coffee.

PART III: SHOULD YOU EAT THAT CHEETO OR NOT?

"YES."

—A scientific study

"NO."

—Another scientific study

CHAPTER SIX

IS COFFEE THE ELIXIR OF LIFE OR THE BLOOD OF THE DEVIL?

This chapter is about coffee, cookbooks, tapioca pudding, french fries, and crumbling cookies.

If you happened to be alive and consuming the news in the mid-1980s, you would have thought that coffee was terrible for you:

COFFEE DRINKING LINKED TO HEART DISEASE RISKS IN WOMEN

LUNG CANCER RISK 'POSSIBLE FROM COFFEE'

FIVE CUPS OF COFFEE TRIPLE RISK

STUDY: COFFEE DRINKERS FACE INCREASED CANCER RISK

STUDY SAYS COFFEE MAY DOUBLE RISK OF HEART DISEASE

But then the Associated Press published the following in early 1987:

COFFEE DOES NOT INCREASE HEART DISEASE RISK, STUDY FINDS

Phew. But then, just two years later, in 1989:

DECAFFEINATED COFFEE MAY BE RISK*

And the scary headlines continued in 1990:

EVEN TWO CUPS OF COFFEE BOOST DEATH RISK

COFFEE PUTS HEART AT RISK

That last one was published on September 14, 1990. And then, *literally twenty-eight days later:*

COFFEE NOT HEART RISK

HEART RISK, COFFEE LINK DISSOLVES

COFFEE POSES NO RISK TO HEART, STUDY SAYS

....
* Et tu, decaf???

But six months after that:

COFFEE TIED TO INCREASE IN HEART RISK

One year later, things finally seemed to settle down:

COFFEE DOESN'T INCREASE RISK OF HEART DISEASE

STUDY SAYS 3 COFFEES DAILY NO RISK TO FETUS

STUDY SAYS COFFEE DOES NOT INCREASE RISK OF BLADDER CANCER

COFFEE NO RISK SAYS STUDY

You'd think this little game would have been over by then, but no. Only twenty-two days after that last headline was published, coffee was back to killing you:

STUDY: HEAVY COFFEE DRINKERS AT GREATER HEART ATTACK RISK

But after twenty-five years of vacillating between HOLY SHIT DON'T DRINK COFFEE or EH IT'S PROBABLY FINE, coffee decided to hit us with a hard key change:

COFFEE DRINKERS DIMINISH HEART ATTACK RISKS: RESEARCHERS

Wait, what? Was coffee actually *good* for you!? The headlines over the next few years would not have helped you make up your mind:

DAMNIT! DRINK LESS COFFEE AND WALK TO LOWER HIP-FRACTURE RISK

YAY! COFFEE CAN CUT CANCER RISK: STUDY

DAMNIT! HEAVY COFFEE CONSUMPTION INCREASES RISK OF AMI IN WOMEN

YAY! COFFEE NOT SIGNIFICANT RISK FACTOR FOR CHD IN AMERICAN WOMEN

YAY! DRINKING COFFEE MAY REDUCE SUICIDE RISK, RESEARCH HINTS

DAMNIT! HEAVY COFFEE DRINKERS RISK DEVELOPMENT OF HYPERTENSION

DAMNIT! CHOLESTEROL RISK MAY BE BREWING IN COFFEE CUPS

YAY! COFFEE MAY CUT COLON CANCER RISK

YAY! COFFEE REDUCES GALLSTONE RISK

DAMNIT! COFFEE, TEA LINKED TO HEART DISEASE RISK IN UK

YAY! COFFEE, NOT TEA, ASSOCIATED WITH LOWER RISK OF CHD

All of these headlines are from *before the year 2000*. After Y2K, the pace of their publication accelerated. In a highly nonscientific mini-experiment, I searched Lexis Nexis for news articles between 2000 and 2019 in health sections of newspapers and wire services that had "coffee," "risk," and "increase" or "decrease" in the story; I got 2,475 hits for "increase" and 615 hits for "decrease." Even if you assume that half of these results are not relevant and half of the remaining ones cover the same underlying science, that's still 600-plus stories telling you coffee increases your risk of something and 150-plus stories telling you it decreases your risk of something.

My initial reaction to this was:

Are
  You
    Kidding
     Me
      This
     Is
    A
     National
      Embarrassment.

Come on, science! It's a simple question: Is coffee good or bad? Should I drink it? Look, I know research is hard, but isn't MORE THAN TWENTY YEARS enough time to come up with an answer?

Coffee is far from the only food that generates conflicting headlines. In 2016, two scientists at Stanford Medical School pulled *The Boston Cooking-School Cook Book* off the shelf and chose fifty ingredients at random, then trawled the literature for studies that attempted to link each ingredient to cancer. (These were not obscure ingredients like sweat from the udder of a lactating goat milked during a full moon. These were salt-of-the-earth ingredients like eggs, bread, butter, lemon, carrot, milk, bacon, and rum.) After excluding ingredients for which there were fewer than ten studies, there were twenty ingredients left. Of those twenty, only *four* had results that entirely agreed with each other. The rest—in other words, 80 percent of the ingredients—had studies with at least one conflicting result. Some, like wine, potatoes, milk, eggs, corn, cheese, butter, and, yep, coffee, had *multiple* conflicting results. And that's how you end up with what statistician and science communicator Regina Nuzzo calls "whiplash news."

We get pissed at politicians when they change their mind *once*; how the hell can science change its mind about a single food dozens of times?

Ladies and gents, I'd like to officially welcome you to—angelic choir sound effect—*nutritional epidemiology.* Nutritional epidemiology is the study of what foods are going to send you to an early grave, and the source of most headlines you read about food and health.

Nutritional epidemiology is mostly based on long-term prospective cohort studies. We've talked about these studies before: they're similar to the ones done in the 1950s on smoking and lung cancer. You enroll a bunch of people, ask them lots of questions about how they live their lives, follow them over a long period of time, and record what diseases they get. What pops out of these studies is an *association* (also called a *correlation*, but I'll stick with the word *association* throughout this book). The studies on smoking found an association between heavy smoking and a 1,000-plus percent increased risk of getting lung cancer. A typical nutritional epidemiology study might find an association between, for example, drinking about two cups of coffee per day and a 30 percent increase in risk of breaking your hip if you fall. This results in the headline:

DRINK LESS COFFEE AND WALK TO LOWER HIP-FRACTURE RISK

Over the years, as more and more nutritional epidemiology studies are done, the associations pile up. Sometimes they're consistent. Sometimes, like with coffee, they're anything but. The associations ping-pong back and forth between good and bad, and health journalists follow each ping and pong, which leads to exactly the type of coffee whiplash I just put you through.

But it wasn't always like this. Let's travel back . . .

. . . through the mists of time . . .

. . . all the way . . .

. . . to 2011.

In 2011, four doctors at the University of Virginia met a patient with right knee pain, which got worse when he put any weight on it. The patient was tired all the time and also had stomach pain, vomiting, diarrhea, and the occasional fever. His right thigh was bruised. Blood tests revealed elevated uric acid levels, and an MRI of the patient's lower half made the doctors suspect the worst: leukemia. To confirm a leukemia diagnosis, you stick a needle into a patient's bone (usually hip) and suck up some of their marrow, which is one of the rare medical procedures that hurts exactly as much as they say it does. In this particular case, the doctors performed bone marrow biopsies of the patient's hip *and* shin bones. But instead of cancer, they found something weirder: the patient's bone marrow was turning into jelly.

I'm now going to reveal something important: the patient was five years old.

Being young is not, in and of itself, a problematic medical condition, but it tends to be accompanied by a lack of elderly wisdom and knowledge about the basics of living. When the doctors asked this kid what he ate, they discovered that he had eaten only the following foods:

- **pancakes**
- **chicken nuggets**
- **tapioca pudding**
- **french fries**
- **animal crackers**
- **vanilla pudding**
- **wheat pretzels**

For.
Three.
Years.

A diet of seven foods, with nary a fruit, vegetable, leafy green, legume, or indeed any food not in a shade of brown to be seen— for three years. Forget leukemia, it's amazing the kid lived that long.

Can you guess what the five-year-old patient was diagnosed with?

Here's a hint: you have probably heard of this disease before. Think about it and then turn the page.

Scurvy. The boy was diagnosed with scurvy. Now, I know what you're thinking: Isn't that the sailor's disease? Yes, it is. For about 350 years, scurvy was the scourge of the sea. Symptoms started gradually—fatigue, aching joints, painful muscles—but got progressively more terrifying. Splotches of blood pooled beneath the skin. Gums bled easily. Body hairs coiled like snakes. Eventually, the disease would kill; some historians estimate that more than two million sailors died this way between roughly 1500 and 1850.

Sailors (and all humans, for that matter), fruit bats, and guinea pigs belong to a very exclusive and unfortunate chemical club: these three species cannot make their own vitamin C. Vitamin C helps your gut absorb iron atoms by giving them an extra electron each, and helps protect DNA. But one of the most important things vitamin C does is take part in a series of chemical reactions that produce collagen, a rigid triple-helical protein that makes up between a quarter and a third of all protein in your body. With vitamin C, collagen has the firmness of an unripe banana; without vitamin C, it has the firmness of . . . a ripe banana that you've frozen and thawed twenty-six times because you thought you were going to make a smoothie but ended up just eating ice cream. That's what causes most of the classic symptoms of scurvy.

For most of human history, we didn't know any of that. European doctors tried and failed for roughly 350 years to figure out what caused scurvy.* If you've ever read anything about scurvy or the history of medicine, or had even five minutes of medical training, you have almost certainly heard of one of these

....

* My favorite old-timey theory: scurvy was caused by your body not being able to fully digest and then sweat out food particles, because sea air was so moist it would close up your pores. Thus, un-sweated-out food particles would unnaturally build up in your body and cause you to rot from the inside out. How very medieval! Then again, some medieval ideas come back around . . . for example, leeches.

European doctors: a Scottish surgeon by the name of James Lind. Lind looms large over medicine, because what he did aboard a British naval vessel in 1747, with twelve people and at a cost of roughly "free" in today's dollars, is what universities, governments, and the world's largest drug companies spend billions of dollars continuing to do to this very day.

James Lind ran a controlled trial.

On a voyage aboard the HMS *Salisbury*, he took twelve sailors with scurvy and split them up into six groups of two. Each group got a different potential cure: a quart of hard cider, seventy-five drops of sulfuric acid, two spoonfuls of vinegar, half a pint of seawater, the "bigness" of a nutmeg,* or two oranges and a lemon. The fact that Lind was comparing multiple treatments (as opposed to just fervently believing in *one* and using that) was remarkable enough, but he also understood that, to effectively compare them, the treatments needed to start on an even playing field. He made sure to choose twelve sailors whose symptoms were as similar as possible. He put them all in the same part of the ship. And finally, he fed them all the same diet. You can guess what happened next: The sailors given oranges and lemons made a full recovery in six days. Those given the cider got a bit better. The rest stayed the same. One June 17, 1747, the ship reached Plymouth, and the experiment was over.

For quite a long time, nutritional science studied these types of diseases:

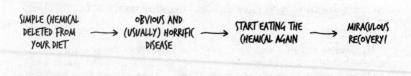

SIMPLE CHEMICAL DELETED FROM YOUR DIET → OBVIOUS AND (USUALLY) HORRIFIC DISEASE → START EATING THE CHEMICAL AGAIN → MIRACULOUS RECOVERY!

....

* This is a weird witches' brew: garlic, mustard seed, radish root, balsam of Peru, and gum myrrh.

Scurvy is the classic example. Vitamin C is made of just twenty atoms, and all you need is 10 milligrams of it per day to avoid a slow and painful death (though to be safe, the Institute of Medicine at the National Academy of Sciences recommends adults get 75 to 90 milligrams daily). Vitamin D is made of seventy-two atoms, and a severe lack of it in kids causes rickets, a painful softening of your bones that can leave you bowlegged and stunt your growth. Vitamin B_1 has thirty-five atoms, and a severe lack of it can cause beriberi, which can cause all kinds of problems in your heart and brain, including death. Pellagra, anemia, goiter, pernicious anemia, xerophthalmia, and many other diseases are all caused by the lack of simple chemicals (in this case vitamin B_3, iron, iodine, vitamin B_{12}, and vitamin A, respectively).

In all these cases, there's an incredibly simple (in hindsight) way to prevent the disease, simply by making sure you get enough of a few different chemicals in your diet.

Want to prevent pellagra? Eat some liver (vitamin B_3, a.k.a. niacin).

Want to prevent goiter? Eat some cod (iodine).

Want to prevent scurvy? Binge on oranges (vitamin C).[*]

In other words, certain foods—and specifically the vitamins and minerals within those foods—were *literally* miracle preventions and even cures for certain diseases. (This is why manufacturers add calcium and vitamin D to milk, or vitamin B_3 to bread: to prevent easily preventable horrific deaths.) These miracle cures and preventions rank right up there with medicine's most effective treatments—its greatest hits, if you will: things like insulin for diabetes and anesthesia for cutting people open. Today we take

....

[*] Obviously, please do not try and WebMD yourself a cure based on this. If you suspect you might have a nutritional deficiency, see a doctor. Many deficiencies these days occur together or occur because of a serious underlying condition. So, yeah, please don't stethoscope yourself.

all this for granted, but it's worth realizing how much of an impact nutritional science had; it basically ended a disease that killed more people than all of America's wars combined—five decades *before* doctors figured out that you should wash your hands after pooping. In the developed world, nutritional deficiency diseases went from killing millions to being medical sideshows.

I'd call that a bed-quaking science orgasm of the first order.

This nutritional sciencegasm taught us a simple relationship:

$$\text{SEVERE LACK OF VITAMIN OR MINERAL} = \text{HORRIBLE, FAST-ACTING, POSSIBLY FATAL DISEASE}$$

This relationship is true—and will remain true. But nutritional deficiency diseases have become rare enough in the U.S., Europe, and the rest of the developed world that the relationship above is no longer relevant for most people. Most of the questions that concern our health these days have nothing to do with scurvy or pellagra and everything to do with heart disease, cancer, diabetes, Alzheimer's, and other chronic diseases. These "new" diseases are very different than scurvy and other "old" ones:

YE OLDE NUTRITIONAL DEFICIENCY DISEASES	YE NEWE NON-NUTRITIONAL-DEFICIENCY DISEASES
Scurvy, pellagra, beriberi	Heart disease, cancer
Happen quickly (months to a few years)	Happen slooooowly (decades)
Happen to all who lack the vitamin or mineral	Happen only to some people
Can happen any time in life	Happen later in life
Horrible, obvious symptoms	Not-so-obvious early symptoms
Quick, dramatic cure	Can be treatable but will eventually kill you

My pet hypothesis, for which I have no proof, is that we're living in the postcoital glow of having discovered how nutritional deficiency diseases work. But our world view hasn't kept up with modern diseases. We've taken the mental framework of the nutritional deficiency diseases:

$$\text{SEVERE LACK OF VITAMIN OR MINERAL} = \text{HORRIBLE, FAST-ACTING, OFTEN FATAL DISEASE}$$

and swapped out a few key words:

SEVERE LACK OF ONIONS = CANCER

or perhaps

ONE TOO MANY CUPS OF COFFEE = HEART ATTACK

or maybe even

SEVERE LACK OF HEMP PROTEIN = ENNUI

Because single foods were miracle cures for horrific nutritional deficiency diseases, we were ready to accept the idea that single foods could be miracle cures for heart disease and cancer. Unfortunately, there are two major challenges associated with Ye Newe Diseases. First, for most of them, you just can't do what James Lind did. Any controlled trial that lasts long enough to determine if a food prevents cancer would be obscenely expensive and basically torture for the participants (no butter . . . *for the rest of your life*). It's *a lot* easier to figure out what causes and cures a fast-acting, dramatically curable disease than a slow-acting one that will probably kill the patient no matter what you throw at it. Remember sunscreen? The same principle applies: it's a lot easier to prevent sunburn—which happens fast and is immediately obvious—than skin cancer.

The second challenge is sort of related. It boils down to this: most health questions we care about these days are not deterministic; they are probabilistic.

What does that mean?

Let's find out.

———

I bet the first chemical reaction you ever saw went something like this. Your parents or elementary school teacher shaped a mound of dirt around a small cylinder, removed the cylinder, put some white powder into the resulting hole, and then poured a clear liquid inside. Immediately a white froth burst from the opening, drenching the sides of the model volcano, and you squealed with chemical delight. That white froth was generated by this reaction:

BAKING SODA + VINEGAR → BUBBLES

If we consider how many ten-year-olds with time on their hands there are on the planet, I think we can say with near certainty that this chemical reaction has been performed millions of times.

Question: Have you ever seen the reaction *not* work? Have you ever mixed the two ingredients and observed . . . nothing? No.[*] Chemical reactions like *baking soda + vinegar* are as simple and reliable as the rising of the sun. If you mix these two chemicals, the bubbles will come. This is what fancy physics folk call "deterministic"; in other words, if I tell you what *is* happening now

....

[*] If you have, it might be because your baking soda was from before there were fifty states, or a prankster replaced your vinegar with water.

(baking soda and vinegar are being mixed), you can tell me what will happen in the future: there will be bubbles.

Sound familiar? If you don't eat enough vitamin C, you *will* get scurvy. The Ye Olde Vitamin Deficiency Diseases are about as close to deterministic as you can get when it comes to food.

But what about this reaction:

HUMAN + CHEETO → ?

What happens if you eat Cheetos or other ultra-processed food? Will human gain weight? Will human get cancer or heart disease? Will human become addicted to Cheetos?

The reaction looks simple, but that's only because we're using simple labels for complex things. Your body, at least as far as we understand it, is thousands of chemical reactions, using up and producing hundreds of billions of molecules. Food, even the ultra-processed kind, is chemically complex enough to interact with your body in ways we can't necessarily predict. And, of course, there are lots of things besides food—like your genes—that affect whether you get a disease or not.

This is what fancy physics folk call "probabilistic": if I tell you what is happening now (human is eating Cheeto), you *cannot* tell me with certainty what will happen in the future. The best you can do is tell me what *might* happen and assign it a probability (for example, there is a 38 percent chance human will get cancer at some point in human's life).

Suppose you walk up to a random person on the street and ask them a simple question like "Is the sky blue?" The answer you get—whether "Yes," "Sometimes," "Fuck off," "Blue," "Purple," or "CATS!"—depends on all kinds of things, including the actual color of the sky at that moment, the person's mood, their willingness to think about the question, their sanity or lack thereof, and lots of other stuff you can't possibly imagine or predict before you ask the question. Ye Newe Chronic Diseases are similar: they depend on many factors, both seen and unseen. Probabilistic diseases are all about risks: if you smoke, you dramatically increase your *risk* of getting lung cancer, but you don't *guarantee* that you'll get it.

Maybe one day we'll have a detailed chemical map of every single person's body and every single food, and maybe that will allow us to predict what will happen to any given person with the same level of certainty that we can predict what will happen when baking soda and vinegar are mixed. But if that ever does happen, it'll be long after we're all dead. If we stick firmly with what's possible in the present, the unfortunate truth is this: the answer to almost every big-picture question that involves chemicals and the human body—Does ultra-processed food give you cancer? Does coffee make you live longer? Does sunscreen prevent skin cancer?—is somewhere between "Maybe" and "Maybe not." Rarely, the results of studies smack you in the face, like with smoking. But more often than not, results are modest and uncertain, like with sunscreen.

And that raises the question: *How do scientists evaluate modest and uncertain results?* And just as important: *Should you change what you put in (and on) your body, based on these results?*

Way back in chapter one, we looked at a few headline-inducing associations between ultra-processed foods and diseases. Ultra-processed food was associated with a higher risk of irritable bowel syndrome, obesity, cancer, and death. Compared to the association between smoking and lung cancer, the ultra-processed food associations are not nearly as dramatic. But that doesn't mean they don't deserve the same loving scrutiny that cigarettes got back in the 1960s.

So we should ask the same questions that scientists asked back then, but this time I'll boil them down to their essence. Any time you read about an association between two things, two questions should pop into your head:

Is this association legit?

If the two things *are* legitimately associated, the next logical question is: Does one of the things *cause* the other? In other words: Is this association causal?

For example: Are ultra-processed foods *legit* associated with cancer?

If they are, does eating a bunch of ultra-processed foods *cause* cancer?

To answer these questions, we have to probe their many subtleties. To do that, we have to leave the comforting haze of nutritional science's postcoital glow and return to the real world.

You probably learned science as I did: by learning all about its successes. Now, let's be real here: most of the material gains made by humanity in the past few centuries are almost entirely thanks to science, and if we ever figure out how to rescue this planet from ourselves, that will be largely due to science, too. To answer questions like *What should I eat?* or *What health information should I trust?* you need to understand science, and yet, ironically, the way most of us learn about science is profoundly *un*scientific. Remember your high school chemistry class? It probably involved halfheartedly memorizing the periodic table and—if you were lucky—doing a few "labs," which were typically just a list of chemicals to mix or burn. Learning chemistry like this is like blindly following a recipe: it's not bad—it develops important hand skills, and you end up with a meal—but it doesn't make you a chef. Much more interesting is how the recipe was created: What was tried? What worked? What failed? Why did it fail? Did people learn from the failure?

Unfortunately, most of us learned about the most important science of our time—the Nobel laureates, the classic experiments, the world-changing theories—via the "Follow the Yellow Brick

Recipe" method. In other words, we didn't have to do much thinking. Again, this isn't bad on its own. What's bad is stopping there. To really understand the issues with nutritional epidemiology or any other science, you have to learn how to appreciate its beauty—and its flaws. You have to learn to spot a mistake or logically tear something apart. You have to sniff out alternative explanations or an argument's weakest link. In short, you have to see the best in people, but you also have to be kind of an asshole.

Don't worry, it's really fun.

CHAPTER SEVEN

ASSOCIATIONS, OR THE GRAPES OF MATH

This chapter is about Ents, private jets, potholes, olive oil, Scorpios, and Santa.

First up on the Be Kind of an Asshole Tour of Science is a deep dive into question one: Are these two things *legit* associated? To be honest, I had given this exactly zero seconds of thought until recently; I simply assumed that if the people who did the study were scientists with fancy titles at Ivy League schools, then the association was legit.

Turns out, that was naïve. Even the most Ivy-Leagued of associations can, upon closer inspection, turn out to be . . . not legit. But what exactly does "not legit" mean? Unfortunately, there isn't a single definition. Instead, allow me to offer an unnecessary analogy: producing a legit association is like driving through a road chock-full of potholes (and prone to the occasional earthquake) and managing to not mess up your car. To see why that's so hard, it's actually easier to explain the potholes than the road. So let's whip out the magnifying class and Sherlock the shit outta these potholes.

First pothole: fraud. No magnifying glass required: scientists can literally make shit up and publish it. Luckily, this is rare.

Second pothole: basic mathematical screw-ups.

Believe it or not, there are basic arithmetic errors in peer-reviewed, published scientific papers. If, for example, you were to pick up the paper "The acute and long-term effects of intracoronary Stem cell Transplantation in 191 patients with chronic heARt failure: the STAR-heart study" and look at Table 2, you'd notice the following calculation:

$$1,539 - 1,546 = -29.3$$

As you'll recall from middle school, when you subtract two whole numbers, the answer *cannot,* in our universe, be a fraction. If you start off with 14 horses and subtract 8, you can't end up with half a horse. Likewise, if you subtract 1,546 from 1,539, you can't have 0.3 in your answer. It's just not possible. And, somewhat more to the point, if you actually subtract those two numbers, the real answer is −7 . . . not −29.3.

Other mistakes are slightly more subtle but no less wrong. Consider this: if a group has 200 patients in it, and you calculate the percentage of patients who have some condition, it is mathematically impossible to get 18.1 percent, which is what appears in Table 1 of that very same paper. Why? Because 18.1 percent of 200 people is 36.2 people . . . 36⅕ people.

Simple arithmetic mistakes are actually the best kind of mistake, because they're fairly easy to spot. As the math gets more complicated, it gets harder to spot the mistakes.

In 2014, three scientists published some highly amazing results in the *World Journal of Acupuncture-Moxibustion.* In a randomized controlled trial, the researchers compared two groups of overweight or obese patients trying to lose weight. One group received meridian massage; the other didn't.* The no-massage

....
* Meridian massage is "a traditional practice that manually stimulates the body's meridian system—the same network of vital energy channels used in acupuncture."

group lost 8.2 pounds in two months; the massage group lost almost double that, 15.4 pounds (more than 9 percent of their baseline body weight), in the same period of time. Losing 10 percent of your body weight in two months would be incredible. To obesity researcher and mathematician Diana Thomas, the results were *literally* not credible. In a letter to the editor of *World Journal of Acupuncture-Moxibustion*, she and her colleagues wrote, "We noticed several oddities," which is science-speak for *You must have been high when you wrote this paper.*

The team that published their original research did not publish their raw data. But they published enough for Thomas to do some mathematical fact-checking. It worked like this: She estimated the change in the two groups' average *height* before and after the treatment. (If you know weight and BMI, you can calculate height.) Everyone enrolled in the study was a human adult, so you would expect that their change in height over a two-month period would be roughly zero. Thomas and her colleagues found that both groups *grew* over the course of the study: the no-massage group by roughly an inch and the massage group by *two and a quarter* inches. So the participants who got massaged lost almost 10 percent of their body weight and also grew 2.25 inches. What could explain these results?

1. **The researchers made them up.**

2. **A few of the participants secretly traveled to Middle-earth, made friends with the Ents, drank a bunch of Ent-draughts, and then traveled back to our world.**

3. **A few of the shorter people dropped out of the study midway through, and the scientists did not correct for this.**

4. **A wide variety of math mistakes.**

Who knows which of these actually happened, but even without looking at the raw data, we know that mistakes were made. It's like if you were wandering around New York City and saw an ostrich trying to steal a Christmas elf from Macy's: you're not sure what went wrong, but you know something definitely did. As I write this, the authors of the original study have not responded to Thomas, and the journal has not retracted the article. (By the way, my money's on mistake number 3.)

Third pothole: procedural mistakes. Just like you can bake a disgusting cake by using a bad recipe or accidentally adding salt instead of sugar, you can destroy a study by planning or executing it poorly. Simple mistakes can be devastating; for example, in a recent study that linked personality traits with political attitudes, researchers accidentally switched the variables for "conservative" and "liberal," so all the associations they reported were . . . exactly the 100 percent *polar opposite* of the actual associations. Instead of confirming, say, the typical finding that people who scored highly on a test for Eysenck's P—which is associated with tough-mindedness and authoritarianism—had politically conservative views about the military, the researchers wrote, "opposite our expectations, higher P scores correlate with more *liberal* military attitudes . . ." So, yeah . . . not great.[*]

Procedural mistakes can also be much more complicated. Let's talk about the PREDIMED trial. PREDIMED, which stands for PREvención con DIeta MEDiterránea (Prevention with Mediterranean Diet), was supposed to definitively answer the question of whether the Mediterranean diet reduces your risk of heart disease. (For those of you who don't remember, the Mediterranean diet was cool before the keto diet. You basically eat plants drenched in olive

....

[*] I will point out, though, that the study wasn't intended to list a bunch of personality traits that were different in conservatives and liberals. It was actually intended to try and figure out whether personality traits *caused* political attitudes (or the other way around).

oil, with the occasional fish and glass of red wine thrown in.) PRE-DIMED was a huge long-term randomized controlled trial: it had almost 8,000 participants followed up for five years. It probably cost more than a Gulfstream G650 (colloquially known as a "G6"). And it seemed that the money was well spent: the main headline finding, published in 2013, was that people who ate a Mediterranean diet supplemented with olive oil or nuts had a roughly 30 percent lower risk of major cardiovascular events.

Unfortunately, the researchers made a pretty damn big mistake at one of the study centers. Instead of randomizing *people* they randomized *clinics*. In other words, instead of splitting everyone in the village into two groups—the regular diet group and the Mediterranean diet group—they just threw everyone in each clinic into the same group. To understand why that's a big deal, let's assume each clinic serves one village and focus on a single village. Suppose that the village in question happened to sit right on top of a nuclear-powered alien spaceship, and the reactor core of the spaceship was leaking radioactive waste, which increased everyone in the village's risk of having a heart attack by a bajillion percent. Everyone in this poor village is having heart attacks left and right.

Now suppose a group of well-meaning researchers comes along and dumps this entire heart-attack-prone village into the Mediterranean diet group. What would happen? That group's risk of heart attack would skyrocket, and—if you didn't know about the spaceship—it would look like the Mediterranean diet was *causing* all these heart attacks. Or if the researchers dumped the entire village into the control (regular diet) group, *its* risk of heart attack would skyrocket relative to the Mediterranean diet group's. And that would make the Mediterranean diet seem like a miracle cure.

Obviously, there are no alien spaceships hiding under Spanish villages (that I'm aware of). But the point is that people who live close to each other can all be exposed to things that improve or worsen their health. And if you don't randomize those people, you can artificially inflate or deflate the effectiveness of whatever drug, diet, or other intervention you're testing.[*]

PREDIMED's mistake was discovered *five years* after the study was initially published. The *New England Journal of Medicine* retracted the paper but allowed the authors to reanalyze the data (excluding the nonrandomized village) and republish the study. Perhaps unsurprisingly, the authors reached basically the same conclusion as before. The data isn't public, though, so a few of the epidemiologists I spoke to are still skeptical of the results. Whether you believe the final result or not, nobody—not even the original authors—disputes that not properly randomizing people in one of the villages was a mistake.

Maybe it was idealistic and naïve of me to expect that the scientific literature would be completely free of silly mathematical and procedural mistakes. Then again, scientists are human, so I guess I shouldn't have been surprised. Either way, the important question isn't *whether* mistakes exist in the literature; it's *how many* of them there are, and how big they are.

Unfortunately, that's really hard to tell. Basically, the only way to find out about a mistake in a paper is if other scientists point out the mistake—publicly. And that's not a pleasant experience for anyone involved. Calling out an error in a scientific journal is like forcing your way into the kitchen of a two-star Michelin restaurant and asking the chef to remake the *sylphides à la crème d'écrevisses* in front of you (and everyone else in the restaurant) to make sure it really is gluten-free. It's annoying for you, humiliating to the chef, and only ends badly for either or both parties.

But some scientists seem to have no problem doing this. In the course of my research, whenever I stumbled across a paper

....

* Confusingly, randomizing everyone across all villages has its own problems. For example, sometimes people share the treatment with a neighbor who's supposed to be in the control group. This would underestimate the effect of the treatment.

calling out errors in another paper, one of the authors was usually David Allison, a longtime obesity researcher. I called him to ask if he could give me an estimate of how many mistakes there are in the scientific literature. He answered me with a metaphor:

> If you were to ask me, "Are there a lot of cracks in the sidewalks in most cities?" I would say, "Well, I've never done a formal analysis, but I can tell you that I often go out walking, and every time I walk more than ten minutes, I see at least one crack in the sidewalk. So it seems to me like there's probably very many cracks in sidewalks." And that's sort of what I can say about the literature. Every time I go for a walk in the literature, I find a few papers that are clearly and unequivocally incorrect.

So there you go.

Okay, three potholes down; four to go.

The fourth pothole on the road to a legitimate association is Random Chance. To explain this, let's pick on some Canadians (eh!). First, a somewhat surprising fact: most residents of Ontario, Canada, are in a gigantic database with the quintessentially no-nonsense Canadian name the Registered Persons Database, or RPDB. This database includes basic information (name, date of birth, etc.) on more than 10 million Ontarians, but its real power is that everyone is assigned a unique ID number. Whenever an Ontarian goes to the hospital, every single treatment that person receives is recorded in another database—using that same ID number.

These data are not public, but researchers can apply to get access to an anonymized version of it, which they can use to answer important questions, like "Do people use more public health care as they get older?" But they can also answer other, less important questions, like "Are Geminis more likely to be alcoholics?" and "Are Virgos more likely to vomit excessively during pregnancy?" Look carefully at those questions and you'll see our old friends—associations—in disguise. Asking, "Are Geminis more likely to be alcoholics?" is the same as asking, "Is being a Gemini *associated* with an increased risk of alcoholism?" Leaving questions like these unanswered would be a crime against science, so, back in the mid-2000s, a group of researchers set out to answer them. The team, led by Peter Austin, got access to the databases and generated comparisons that looked like this:

	GEMINI	ALL OTHER SIGNS
Percentage of people hospitalized in the year after their birthday (in 2000) for alcohol dependence syndrome	0.61%	0.47%

Translation: If you were a Gemini living in Ontario in the year 2000, your risk of being admitted to the hospital for alcoholism was 0.61 percent. If you were any other sign, your risk was 0.47 percent. So Geminis were 30 percent more likely (0.61/0.47 = 130 percent) to be admitted for alcoholism than the average of all other astrological signs.[*] In the language of associations: Austin

....

[*] Unfortunately, the original paper didn't include the actual percentages for every sign, so I used the U.S. alcohol dependence statistics to come up

found that there was an association between being a Gemini and a 30 percent higher risk of admission to the hospital for alcoholism. But was that association *legit*?

Let's consult our Pothole-O-Meter.

First, let's assume that Austin and colleagues didn't do any fraud or make any basic math mistakes.

Potholes #1 and #2 avoided: ✓

Let's also assume that the study was procedurally sound; that, for example, Geminis weren't mislabeled as Virgos, or that doctors didn't misdiagnose Geminis more or less than other signs.

Pothole #3 avoided: ✓

All right, so if the hospitalization numbers are untainted by fraud, math mistakes, or procedural errors, then the association between Geminis and alcoholism is legit. Right?

Maybe.

There's something else that might have caused the association between being a Gemini and alcoholism: Random Chance. I find this phrase highly unsatisfying, because Random Chance is not a clear and defined cause. Instead, it's more like . . . well, imagine if you were to take a crispy Chips Ahoy!–type cookie and crumble it up in your hands, letting the crumbs fall a few feet to the floor. Then you move to a different spot and do it again with another cookie. And again. You could repeat this experiment a million times and never get exactly the same pattern of crumbs on the floor. Even though your hands and the cookie are subject to the laws of physics, the cookie never crumbles exactly the same way twice. Random Chance is pretty much that: the way the cookie crumbles.

And as psychologist Brian Nosek points out, "Randomness can produce things that look real." In other words, sometimes those

....

with at least semi-plausible numbers. The ratio between the two numbers (130 percent) is accurate and in the original paper.

crumbs are going to look like Jesus. Or in this case, associations between an astrological sign and alcoholism.

So the question is: How do you tell if an association is caused by Random Chance? *Can* you even tell?

This is where things get really hairy. There's a branch of mathematics (created as Lucifer fell from the heavens) called "inferential statistics." There are many different tools in this kit, but by far the most popular is a calculation that produces what's known as a "p-value." This p-value is a number ranging from 0 to 1. To explain what it means, let's use our friends the Geminis as an example. Austin and colleagues calculated a p-value for this Gemini-vs.-non-Gemini difference: it was 0.015.

What does this mean? Here's an accurate definition:

> *The p-value is the probability that if you were to compare a group of randomly selected Geminis to another group of randomly selected non-Geminis, the difference in alcoholism between the two groups would be at least as high as what Austin found (0.14 percentage points), if these three things were true: (1) across everybody in the entire universe, there is truly no difference in alcoholism between Geminis and non-Geminis, (2) all assumptions made in the construction of Austin's statistico-mathematical models were valid, and (3) every single step of Austin's study was unblemished by fraud, math mistakes, procedural errors, other potholes we haven't talked about yet, or any other chicanery, tomfoolery, hogwash, poppycock, or drivel.*

This definition is an unholy mess. So most scientists, journalists, policy makers, and pretty much everyone except professional statisticians simply ignored it and pretended that this was the definition of a p-value instead:

*The p-value is the probability that Random Chance caused
the association between Geminis and alcoholism.*

If you go by the second (pretend) definition, you might look at
the p-value of 0.015 and conclude that:

1. **There is only a 1.5 percent chance that
   Random Chance caused the association
   between Geminis and alcoholism;**

2. **Therefore, there is a 100 − 1.5 = 98.5 percent
   chance that Random Chance did not cause
   the association;**

3. **Therefore, there is a 98.5 percent chance the
   association is legit.**

For a long time, many scientists operated using this mental
framework. They agreed that if the p-value was below 0.05 (5
percent), then the association would be anointed as "statistically
significant" and considered legit. If a p-value was—*gasp!*—over
0.05, the result was "not statistically significant" and considered not
legit. This wasn't just an academic difference: if you were a professional scientist, your *job* was to publish statistically significant papers. If you did, you got tenure. If you didn't, you started a bakery.

Unfortunately, using a p-value to decide what's legit is wronger
than brie on borscht.

If you look back at the *accurate* definition of a p-value, you'll
notice that the second or third assumptions can be violated for
almost any reason. Austin's p-value of 0.015 could have been
caused by anti-Canadian hackers maliciously changing numbers
in Canadian databases, or because Austin divided some number
instead of multiplying it, or because doctors classified more Geminis as alcoholics, or for hundreds of other reasons.

Probably the simplest way to think about a p-value is, as Regina Nuzzo says, a "measure of surprise." Picture this: it's 2:00 A.M. on Christmas and you're woken up by a noise in the living room. *Holy shit!* you think to yourself. *IT'S* SANTA*!*

Or is it?

It certainly *could* be Santa. There's no law of physics barring his existence. But it could also be your kid sneaking downstairs, hoping to get a peek at Santa. Or your thirty-six-year-old brother eating Santa's cookies. Or a book falling off the shelf. Or a thief breaking in. A low p-value is like a sound in the night: it tells you *something* unexpected is going on, but it doesn't tell you *what* that something is. Just because the noise is loud enough that you're 99 percent sure that something happened downstairs does *not* mean you should be 99 percent sure that Santa impaled himself on the fireplace poker.

Let's take stock. Random Chance—the fourth pothole on the road to a legit association—is the most complicated one so far. Unlike the first three, it's not entirely our fault. It's just the way the universe works: sometimes the cookie is going to crumble into something that looks like an association, when really it's just a random fluke. Unlike the other potholes on our list, Random Chance is not fixable: it's just something we have to try and understand. Unfortunately, we've managed to spectacularly misunderstand p-values for decades, and while that's not a pothole on its own, it helps set up the biggest one yet. Let's get back to our alcoholic Geminis.

I've been hiding something from you. Peter Austin and his colleagues didn't *just* find that Geminis are more likely to be hospitalized for alcoholism . . . They found many other associations between astrological signs and diseases. They basically came up with a Scientific Horoscope:

~~~YOUR SCIENTIFIC HOROSCOPE~~~

ARIES	♈	41% MORE LIKELY TO BE HOSPITALIZED FOR INTESTINAL INFECTION DUE TO OTHER ORGANISMS
TAURUS	♉	27% MORE LIKELY TO BE HOSPITALIZED FOR DIVERTICULA OF THE INTESTINE
GEMINI	♊	30% MORE LIKELY TO BE HOSPITALIZED FOR ALCOHOL DEPENDENCE SYNDROME
CANCER	♋	12% MORE LIKELY TO BE HOSPITALIZED FOR INTESTINAL OBSTRUCTIONS (WITHOUT HERNIA)
LEO	♌	17% MORE LIKELY TO BE HOSPITALIZED FOR UNSPECIFIED PROCEDURES
VIRGO	♍	40% MORE LIKELY TO BE HOSPITALIZED FOR EXCESSIVE VOMITING DURING PREGNANCY
LIBRA	♎	37% MORE LIKELY TO BE HOSPITALIZED FOR FRACTURE OF THE PELVIS
SCORPIO	♏	57% MORE LIKELY TO BE HOSPITALIZED FOR ABSCESSES OF THE ANAL AND RECTAL REGION
SAGITTARIUS	♐	28% MORE LIKELY TO BE HOSPITALIZED FOR FRACTURE OF THE HUMERUS
CAPRICORN	♑	29% MORE LIKELY TO BE HOSPITALIZED FOR OTHER ILL-DEFINED AND UNKNOWN CAUSES
AQUARIUS	♒	23% MORE LIKELY TO BE HOSPITALIZED FOR CHEST PAIN
PISCES	♓	13% MORE LIKELY TO BE HOSPITALIZED FOR HEART FAILURE

In total, they came up with seventy-two diagnoses like the ones above, in which one particular sign was associated with a statistically significantly higher chance of being hospitalized than all the others combined. All the p-values for all the associations were less than 0.05, a.k.a. "statistically significant." So Austin et al. concluded that all the seventy-two associations they found were legit. To all my fellow Scorpios: astrology is real—enjoy your anal abscesses!

JUST.

KIDDING.

I've been describing this as if it were real science. And, on the surface, it is. Austin and colleagues really did all the things they said they did: they searched this massive database, ran all the numbers, and found all the associations listed above (and many more). In that sense, it's real. But Peter Austin is not an astrologer, shaman, or medical doctor. He's a statistician. And his experiment was a demonstration of how blindly following the wrong mental framework can yield . . . a whole bunch of Santas. In sum: this experiment was a statistical crash test dummy, meant to show the perils of Pothole #5: p-hacking, also known as fiddle-assing with data until you "find" what you were looking for.

Let's replay this crash in slow motion. There are two critical mistakes.

First, Austin adopted the convention that if p is less than 0.05, the association is legit. This is dead wrong. There is actually *no* p-value that guarantees that an association is legit. The p-value is a clue, but it is rarely the most important one, and it's definitely not The Great Revealer of Fundamental Truths. It's a sound in the night, not conclusive evidence that Santa exists.

The second critical mistake is this: Austin and colleagues cast an experimental net wider than a liberal's Solstice Inclusivity Gathering. Instead of formulating a single specific hypothesis about a single astrological sign or diagnosis, they formulated and then tested 14,718 hypotheses! All it took was a gigantic database and some code to make thousands of comparisons, asking very similar questions over and over again (see opposite).

Every single one of these questions is its own experiment. So Austin wasn't running *one* experiment; he was running over 14,000.[*]

....

[*] You might be wondering why there were only seventy-two statistically significant associations. After all, Austin ran about 14,000 experiments and the threshold p-value was 0.05, so you'd expect 14,000 × 0.05 = 700 to pop up as statistically significant. What gives? Austin wrote about only seventy-two of the associations. There were undoubtedly others hidden in the data, but unfortunately, he didn't compile a full list. Still, seventy-two astrologically linked medical "diagnoses" is more than enough, thank you very much.

ARE VIRGOS ♍ SIGNIFICANTLY MORE LIKELY TO BE HOSPITALIZED FOR TUBERCULOSIS?

WHAT ABOUT SYPHILIS?

WHAT ABOUT GOUT?

WHAT ABOUT APPENDICITIS?

WHAT ABOUT... (ETC ETC ETC ETC ETC)

ARE LIBRAS ♎ SIGNIFICANTLY MORE LIKELY TO BE HOSPITALIZED FOR TUBERCULOSIS?

WHAT ABOUT SYPHILIS?

WHAT ABOUT GOUT?

WHAT ABOUT APPENDICITIS?

WHAT ABOUT... (ETC ETC ETC ETC ETC)

What's so bad about that? Doing what Austin and colleagues did—casting a way-too-wide experimental net and then choosing the results that happen to be significant—is like having five kids, waiting thirty years to see which ones are the most successful (p < 0.05), disowning the others (p > 0.05), and then declaring yourself the best parent in the history of parenting (publishing only the p < 0.05 experiment). Austin could have taken this giant data set, run his 14,000-plus experiments, "discovered" that Geminis are 30 percent more likely to be hospitalized for alcoholism than other signs, and published *only that result*.

The more kids you have, the more likely it is that at least one of them will be successful—whether you're a good parent or not. Likewise, the more hypotheses you test, the more likely at least one of them will be "statistically significant" simply by Random Chance.

What we just talked about is the bluntest form of p-hacking: testing thousands of hypotheses and publishing only the ones with $p < 0.05$. There are far more subtle ways to do it—ways that even professional scientists may not recognize as hacking. Let's do a quick thought experiment. Suppose that instead of running 14,000 experiments, Austin ran only one: he had a theory that Scorpios are more likely to be alcoholics, so he scanned the database for *only* that. And he found that Scorpios are 37 percent more likely to be alcoholics! But alas, the p-value was 0.76 . . . way above 0.05 and therefore not publishable. So does he just give up and move on to something else?

Hell no.

He's a scientist: he's spent his entire life turning lemons into lemonade, succeeding, being undaunted by failure. He's not going to just *give up*. As if.

Instead, he might say to himself, *You know, this data is only for the year 2000. Maybe if I combine with data from 1999 and try again, I'll find something.*

So he does. Result? P-value of 0.43.

Okay, now he's getting somewhere. So he tries using *only* 1999 data.

$p = 0.12$

Ooh, tantalizingly close!

Then a thought strikes him: children can't be alcoholics (one would hope). So maybe he should try again, this time using data only from people over the age of eighteen.

$p = 0.071$

Almost there!

Now that he's thinking about it, maybe eighteen is the wrong cutoff. Perhaps the pull of the Mercury is strongest in your thirties, so he tries again, this time using data only from thirty- to forty-year-olds.

$p = 0.98$

Dagnabbit!

Then yet another thought strikes him. It's probably unusual

to have alcohol dependence syndrome in college. So he tries again, this time with using data only from people over the age of twenty-two.

p = 0.043

Jackpot! Publish!

What Austin just did (in our thought experiment) is a more subtle form of p-hacking. Instead of running thousands of experiments, he ran one and then tweaked it until he got what he was looking for. In this example, he manipulated only a couple variables: people's age and the year they were hospitalized. But he could also have added more people from different cities, split up the data by sex, tweaked the particulars of the algorithm he was using to calculate the associations, or literally hundreds of other data manipulations.

What makes p-hacking so insidious is that it seems so . . . good, like you're persevering in the face of reluctant data, until you find *THE TRUTH*. As three psychologists put it in a recent review paper: p-hacking "is not something that malevolent researchers engage in while laughing maniacally; it is something that benevolent researchers engage in while trying to understand their otherwise imperfect results."

And some would say it's about much more than just trying to understand results. Many of the researchers I talked to laid some of the blame on the powerful pressure to publish "statistically significant" results. Regina Nuzzo said it best:

> *We've set up the reward system such that you're supposed to achieve statistical significance. Like you're achieving orgasm or something. That's what they make it sound like, right? Like you have to keep going until you reach the climax!*

But, she adds, "that's not the way it should be, either in sex or in science. It's the process that matters."

So. Quick recap of the potholes on the road to a legit association:

Pothole #1: fraud

Pothole #2: basic mathematical mess-ups

Pothole #3: procedural problems

Pothole #4: Random Chance

Pothole #5: statistical skulduggery, including p-hacking

Now let's review some of the terrifying numbers from chapter one:

EATING FOUR TIMES AS MUCH ULTRA-PROCESSED FOOD	26% HIGHER RISK OF BECOMING OVERWEIGHT OR OBESE
EATING 2.5 TIMES AS MUCH ULTRA-PROCESSED FOOD	21% HIGHER RISK OF DEVELOPING HIGH BLOOD PRESSURE
EATING FOUR TIMES AS MUCH ULTRA-PROCESSED FOOD	23% HIGHER RISK OF GETTING CANCER
EATING TWICE AS MUCH ULTRA-PROCESSED FOOD	25% HIGHER RISK OF IRRITABLE BOWEL SYNDROME
EATING 10% MORE ULTRA-PROCESSED FOOD	14% HIGHER RISK OF DEATH

All these associations come from two large cohort studies: one in Spain, called the Seguimiento University de Navarra, or "SUN" study, and the other in France, called "NutriNet-Santé" study.

Okay, now let's talk about the potholes.

Pothole #1: fraud

Let's assume there wasn't any.

Pothole #2: basic mathematical mess-ups

Let's also assume there are zero silly math mistakes in any of these studies. We kinda have to make this assumption, even though it's probably too generous, because we don't have access to the raw data. So when a paper says "21 percent increase in risk," we're going to take that number at face value and assume the authors didn't accidentally mistype "12 percent increase in risk."

Pothole #3: procedural problems

Next let's look at what exactly the data is. Both of these large cohort studies are essentially gigantic ongoing surveys. The main data collection method is either to mail people hard copy surveys or to have them fill them out on the Internet. That means both of these studies rely on people accurately remembering (and truthfully telling you) all kinds of things: what they ate, whether they're pregnant, how much they weigh, how tall they are, what their cholesterol levels are, and on and on for hundreds of items. The SUN study asked participants 554 questions when they first joined the study, and a couple hundred more every two years thereafter. From what I can tell, it seems like almost all measurements in these studies were self-reported. In other words, at no point did a nurse, doctor, scientist, or anyone else actually meet, let alone draw blood, measure, weigh, poke, or prod the vast majority of the participants. Only some of the participants were actually measured in person to compare against their self-reported measurements; most just filled out surveys.

Even if people weren't lying and had 100 percent accurate memories, the food surveys used in these studies capture only a few moments in time. Data from the French study that linked ultra-processed food consumption with death came from people who filled out an average of six food questionnaires (each covering a twenty-four-hour period) over a span of two years. If you happened to catch someone right after they went to a one-year-old's birthday party, you might get a wild overestimate of how much ultra-processed food they ate; if they're in the middle of a juice cleanse, you might get a wild underestimate. This type of error could lead you astray in two ways: either overestimating or underestimating risk. In other words, it could be a false alarm or an alarm that doesn't go off when it should.

These types of food surveys inspire some of the most heated debates I've ever seen in science—but we'll come back to that later.

Pothole #4: Random Chance

Coulda been a factor. But as we've seen, you can't just look at a p-value and know whether Random Chance caused a result. But there is something else you can do: chill. Hang. Relax. Basically: wait. Why? To see if other scientists replicate or refute the result. As I write this, two more studies have just come out linking ultra-processed food to bad outcomes—one of which is death. But it's still early days.

Pothole #5: statistical skulduggery, including p-hacking

Large prospective cohort studies on nutrition usually measure hundreds of variables (height, weight, blood type, education level, quantity of fish eaten per day, number of bags of Cheetos eaten per day, etc., etc., ad infinitum), and there are hundreds of choices that investigators can make when analyzing the data (who to include/exclude, how long the follow-up should be, what mathematical model to use, etc.). In other words, scientists have a *ton* of options on how to build their study; and *that* means that p-hacking—whether conscious or not—is much easier. Unfortunately, it's pretty much impossible to read a study and be able to tell whether it was p-hacked or not . . . unless the professor responsible happens to write a long blog post in which he unknowingly admits to making a grad student p-hack her results. (Yep, this actually happened. Google "Brian Wansink.")

That having been said . . .

When you read a result that comes out of one of these large prospective cohort studies, imagine this scenario: You're at the neighborhood July 4th barbecue: there are burgers, dogs, and all the families with their teenage kids. The parents hosting the event introduce you to their daughter, who got straight A's and is now midway through a summer internship at Witherspoon Fenway & Creek LLP; and you think, *Wow, these parents must be awesome parents!* But here's the thing: there's no guarantee that *all* their kids showed up. There might be a secret brother who's tired of going to the July 4th party every year and is currently holed up in his room, huffing paint and sending netherpics to his teachers. In other words, you may be seeing *only* the set of variables and analysis that resulted in a "successful" association.

I ran the kid-huffing-paint metaphor by Brian Nosek, and he was really weirded out. To his credit, instead of hanging up the phone, he offered an equally valid but much less weird metric: "When you can say in advance, 'This is what I'm intending to do. This is what I'm predicting. This is what I think's going to happen next,' then I'll be impressed. After the fact—not so impressive."

Let's look at a specific example.

The NutriNet-Santé study tested associations between ultra-processed food and six cancer outcomes: prostate, colorectal, breast, premenopausal breast, postmenopausal breast, and all cancers.

Or . . . did it?

There are more than a hundred different types of cancer.

Was there an association between ultra-processed foods and stomach cancer? Let's imagine the authors tested this hypothesis and found:

p = 0.35

What about esophageal cancer?

p = 0.78

Brain cancer?

p = 0.09

Postmenopausal breast cancer?

p = 0.02

Bingo!

See where I'm going with this? And "type of cancer" is just one variable. There are hundreds of others, both explicit and implicit, that the researchers could have played around with. Look, there's nothing inherently wrong with whittling down one hundred types of cancer to six or choosing any of the other variables. As a scientist, you *have* to make choices about what to test. But I believe that you, the reader of the study, are entitled to a guarantee that the variables were chosen *before* any of the data analysis happened, or at least a disclaimer that they weren't.

Scientists have a fancy name for this kind of guarantee: preregistration.

Preregistration is where you tell the world exactly which variables you're going to test—and exactly how you're going to analyze the data—before you enroll a single participant in your study. If you look up the SUN study and the NutriNet-Santé study in NIH's preregistration database, you will find both of them there.

So . . . box checked?

No.

Both of them were "preregistered" years *after* they had actually started. That's not how preregistration is supposed to work. To be fair, preregistration wasn't really a thing when either of these studies was started, but it had become a thing well before any of the papers specifically on ultra-processed food were published. So, ideally, the authors would have preregistered their data analysis plan, saying, "We want to analyze our data set to see if ultra-processed foods are associated with overweight and obesity [in the case of the SUN study] and with these six cancers [in the case of the NutriNet-Santé study], and here's exactly how we're going to crunch the numbers." As far as I can tell, none of them did that. In fact, none of the preregistration materials for either study mention ultra-processed food at all.

So . . . where does this leave us?

Of the potholes on the road to a legit association we've looked at so far, basic arithmetic and procedural screw-ups are the most fun, because they're unquestionably, certainly, definitely *wrong*. That's why the PREDIMED screw-up made headlines around the world. But the thing that itches in my brain the most—the thing that makes me question all the scary numbers from chapter one the most—is p-hacking, because you can't tell just by reading a study whether the result is a legit association or creative p-hacking.

But hang on. We're getting a bit ahead of the game here. There are other potholes we haven't even talked about yet.

CHAPTER EIGHT

WHAT'S THAT PUBLIC POOL SMELL MADE OF?

This chapter is about coffee (again), chlorine, public pools, red underwear, and quesadillas.

The potholes we've looked at so far are potholes on the road to a *legit* association. But let's suppose for a moment that you've got an association that is 100 percent absolutely, positively, rock-solidly, legitimately legit. How can you be so sure? Because you got it from a burning bush. The bush said unto you, "Shotgun ownership is very strongly associated with having more female sexual partners." For the sake of argument, let's assume that God doesn't p-hack or make silly math mistakes. So you know the association is legit. Remember from chapter six that the next question you have to ask yourself is:

Is the association *causal*?

In other words, do women prefer to sleep with shotgun owners *because* of the shotgun?

And the whole point of *that* is to answer the obvious follow-up:

If I were to buy a shotgun, would women suddenly start jumping into bed with me?

No.

There's something else I haven't told you about that's causing *both* shotgun ownership and having more female sexual partners.

Guess what it is, then turn the page.

It's checking the box marked "male" on the survey.

When you stop and think about it, that's not surprising. If you identify as male, you're statistically more likely to buy a shotgun and you're also statistically more likely to have sex with women. In the language of associations, the association between shotgun ownership and having more female sexual partners is *legit*—but *not* causal. So if you buy a shotgun for the express purpose of sleeping with more women . . . sorry, that probably won't work, no matter how you identify.

Legitimate but noncausal associations that are caused by some other hidden factor are called confounded associations. Unfortunately, they're usually a lot harder to spot than in the somewhat contrived (but true) example above.

Let's look at a confounded association out in the wild.

Multiple studies over the years have found an association between coffee and an increased risk of lung cancer. One analysis found that coffee drinkers were 28 percent more likely to get lung cancer than non–coffee drinkers. This was based on eight studies that reported more than 11,000 cases of lung cancer, with an overall p-value of 0.004.

It's also a bit strange: How might something that never touches your lungs cause lung cancer? Remember NNK, the potent carcinogen in cigarettes that gives rats lung cancer no matter how you get it into the rats? Maybe coffee contains NNK . . .

Turns out it doesn't, but it does contain a chemical called acrylamide, which is also present in cigarettes and fried starchy foods (among others). The International Agency for Research on Cancer, the U.S. National Toxicology Program, and the U.S. Environmental Protection Agency all say that acrylamide is probably a human carcinogen based on its ability to cause thyroid cancer in rats and mice.

So: acrylamide in coffee causes lung cancer. Case closed?

Nope.

For one thing, the doses of acrylamide that caused cancer in lab animals were 1,000 to 10,000 times higher than what a human would consume in coffee. For another, in addition to containing at least one chemical suspected of *causing* cancer, coffee also contains chemicals suspected of *preventing* cancer. But more important than either of those two things, there's a third, lurking factor: smoking.

As you already know from chapter four, smoking causes a dramatic increase in the risk of lung cancer. And smoking is strongly associated with drinking coffee.

Our original picture of the situation looked like this:

But now we have a slightly more complicated picture:

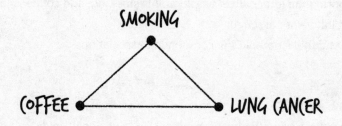

So . . . which thing is flooring the lung cancer accelerator, coffee or smoking? There are three ways to answer this question: easy, medium, and hard. The easy way is to just assume that the most likely reason that coffee (or anything else) is associated with lung cancer is because of smoking's incredibly powerful and causal association with lung cancer. That's not a crazy thought, but it's not convincing by itself. The hard way to answer this question, as you've probably guessed, is by setting up a randomized controlled trial in which you take thousands of people, split them up randomly into two groups, make one group drink coffee and the other abstain, and then see who ends up with lung cancer. In addition to being a pain in the ass, this method is ethically dubious and expensive, and it would take at least a decade to answer the question.

The medium-difficulty way is actually the most ingenious. Can you guess what it is? Think about it and then read on. Remember in chapter four when we talked about the fact most people on the planet don't smoke? What if you could redo the coffee/lung cancer association test *only* with people who have never smoked in their lives?

You can.

Scientists did.

And here's what they found: when looking only at people who've never smoked, drinking coffee was actually associated with a slightly *lower* risk of lung cancer, though this result did not achieve statistical significance. That suggests that the douchebag flooring the lung cancer accelerator is smoking, and coffee is just an innocent passenger.

And that's pothole #6: confounded associations.

———

Are there confounded associations in nutritional epidemiology studies?

Let's try to find out by picking on the NutriNet-Santé study. This was the study that found an association between ultra-processed food and cancer.

At the end of the study, the authors split the participants up—mathematically, not physically—into four groups of roughly equal size based on how much ultra-processed food they ate. People in group one ate the least amount of ultra-processed foods: about 8.5 percent of their diet. We'll call them the Quinoa Enthusiasts.

People in group four ate the most: 32.3 percent of their diet, or almost four times as much candy, cake, soda, Oréos,[*] etc., per day as the Quinoa Enthusiasts. We'll call them the Chemists.

Now here's the crucial thing: when you split up people based on one variable—in this case, ultra-processed food—you also split them up by a whole bunch of *other* variables. There's no way to get around this; it's just true. In this case, the Chemists were different from the Quinoa Enthusiasts in a whole bunch of ways. Specifically, the Chemists were more likely to:

- **be younger**
- **smoke**
- **be taller**[†]
- **be less physically active**
- **eat more**
- **drink less**
- **use birth control**
- **have fewer kids**

So when you're comparing the Chemists and the Quinoa Enthusiasts, you're not comparing *just* people who ate more ultra-processed food to people who ate less ultra-processed food. You're comparing:

younger, taller, less physically active, birth-control-using smokers who don't drink and eat a lot of ultra-processed food

vs.

older, shorter, more physically active, non-birth-control-using nonsmokers who drink and eat much less ultra-processed food.

....

* Disappointingly, they're actually just called Oreos in France. #partyfacts
† Fun fact: taller people are (very slightly) more likely to get cancer. #partyfacts

If this sounds like confounded association nirvana, your instincts are dead-on.

Let's take one specific example. Remember that the whole point of this study was to see whether people who ate more ultra-processed food got more cancer. And the headline result from this study was that the Chemists had a 23 percent higher risk of getting cancer than the Quinoa Enthusiasts. But if you actually look at the raw number of cancer cases among these two groups, you'd be in for a surprise. Among the Chemists, there were 368 cancer cases. Among the Quinoa Enthusiasts there were . . . 712 cancer cases. So the people who ate four times the amount of ultra-processed food got *half* the cancer???? What the eff? Does ultra-processed food *prevent* cancer?!?!

No.

As it turns out, the major confounding variable here is *age*. The Quinoa Enthusiasts were, on average, ten years older than the Chemists. Again, there's no way to physically untangle these differences: once you categorize a large group of people based on one variable, other variables are going to be different, too. And some of these—like age—can have a huge effect on the outcome you care about—like cancer.

So, to answer the question we started with: yes. If you do an observational study and mathematically split people up by one variable, you will also split them up by a bunch of other variables, and that will almost certainly produce at least one confounded association.

In theory, you can "adjust" for all these potentially confounding variables, which basically means "do a bunch of math to try and isolate the effect of the variable you care about," in this case, ultra-processed food. That's how the authors of this study got from the raw numbers (Chemists had 48 percent less cancer than Quinoa Enthusiasts) to the final result (Chemists had 23 percent more cancer than Quinoa Enthusiasts). Unfortunately, there are two problems: First, in order to adjust for things, you have to measure them, and it's well-nigh impossible to make sure you've measured every single variable that matters. Second, and more important, adjusting for variables is tricky, and you can never be certain you've adjusted properly. This can produce a mistake in either direction: you can over- or underestimate the true risk.

So, pothole #6, confounded associations, is a stumbling block between a *legit* association and a *causal* one. In this particular case, the association between ultra-processed food and cancer might be *legit*, but it might not necessarily be *causal*. It could be caused by a confounding variable that the researchers didn't measure (religious service attendance or personality traits, for example), or it could be caused by imperfect adjustment of the variables that the researchers did measure (for example, age or smoking).

———

The last pothole we'll cover—pothole #7—is possibly the most subtle one . . . if potholes can be subtle. Anyway, let's go back to coffee, mostly because there have been many more studies done on coffee than on ultra-processed food.

In 2017, researchers published a mammoth study of studies on coffee. In fact, it was actually a study of studies of studies. In other words, scientists reviewed studies that had previously reviewed results from studies of coffee. Basically it was the *Inception* of coffee studies.

Long story short: the authors mathematically combined the results of hundreds of studies on coffee that included millions of people.

That's a lot of data.

Let's look at what is, IMHO, the most important data point: drinking three cups of coffee per day was associated with a roughly 17 percent lower risk of death from any cause compared to people who drank no coffee. Three cups was the sweet spot, although people who somehow managed to drink *seven* cups a day, Lorelai Gilmore–style, had a 10 percent lower risk of death from any cause than people who drank no coffee. In both cases, these comparisons achieved statistical orgasm, a.k.a. significance, and the p-values were very low.

So let's treat this association as though it came from a burning bush: we'll assume that drinking coffee is legitimately associated with a lower overall risk of death. But, of course, just because drinking coffee and reduced risk of death are associated doesn't necessarily mean that coffee *causes* the reduced risk of death.

To see why those two things are different—and just how different they are—I need you to dredge up the memory of a smell: specifically, that hallmark of an American summer, the Pool Smell. If you've ever spent any time in an indoor public pool, you know the smell I mean. It's tangy, heady, and has vague overtones of Lysol, like if Giada De Laurentiis baked a lemon soufflé in a hospital morgue.

What are the ingredients of this smell?

Let's think. The Pool Smell is pretty *un*usual. You don't smell it when you take a shower, or when you boil water, or when it rains, or if you swim in a lake. You smell the Smell only around public pools—and it's much stronger in an indoor pool than an outdoor one. You also know that public pools are disinfected with chlorine. Finally, you probably know that lakes, rivers, and rain are *not* disinfected with chlorine. So let's review what you know about chlorine and the Pool Smell:

	DISINFECTED WITH CHLORINE?	POOL SMELL?
LAKES	No	No
RIVERS	No	No
RAIN	No	No
TAP WATER AT HOME	Yes	No
PUBLIC OUTDOOR POOL	Yes	Yes, not very strong
PUBLIC INDOOR POOL	Yes	Yes, very strong

The above is essentially an observational study: in exactly the same way as researchers collected a bunch of human samples and observed that coffee and a reduced risk of death are associated, you've collected a bunch of water samples (into your nose) and observed that in almost every case where chlorine is present, the Pool Smell is also present. In other words, chlorine and the Pool Smell are associated, just like coffee and lower risk of death. Notice that the association isn't perfect: your tap water at home is almost certainly disinfected with chlorine, but there's no pool smell. But it's pretty close: you might say chlorine and the Pool Smell are *strongly* associated.

Whenever you read that one thing is associated with another, your mind makes an intuitive leap. And it's not just *your* mind; most minds do this, and they do it so often and so naturally that it's hard to notice. Your mind takes the data in that table above and smoothly glides to the conclusion that one of the two things *must be causing* the other. After all, smells are usually caused *by* something, not the cause *of* something. So your mind gently arrives, without your even noticing it, at the conclusion that the

Pool Smell is caused by chlorine. It makes intuitive sense; in fact, you probably know the smell as the Chlorine Smell.

But is that conclusion true?

To find out, I did a simple experiment.[*]

I put 100 milliliters of filtered water into two empty beakers (each) and smelled both: nothing. Good. Water shouldn't smell like anything. Then I added 0.025 grams of calcium hypochlorite, a common chlorine-based pool disinfectant, to one of the beakers. I stirred it for a few minutes to make sure everything was dissolved, then I sniffed. If chlorine caused the Pool Smell, this beaker should have smelled strongly of the Pool Smell.

It didn't.

Hm. Maybe the reaction takes a while to get going. So I covered both beakers and left them overnight at room temperature.

Next day: no smell.

Well, that's weird. What else could be in pools?

Oh.

No.

Not . . .

Pee?

Jesus. Is it pee?

It's gotta be pee . . .

Only one way to find out.

I redid the experiment, this time with four beakers: one with just water, one with water and calcium hypochlorite, one with water, calcium hypochlorite, and less than a drop of still-warm piss, and one with just water and piss. I covered the beakers and left them overnight.

....

* Many others have done this experiment before me; in the spirit of science, I'm attempting to replicate the results and also add a control or two. For a great example of a previous experiment, see https://www.youtube.com /watch?v=S32y9aYEzzo.

Next morning: sniff.

oh, no

no, no

nonononononono

NOOOOOOOOOOOOOOOOOOOOOOOOOOOOOOOOOOOOOOOOOOOOOOOOOOOOOOOOOO!!!!

Turns out, it's not one or the other; it's both.

Does that mean that all those pools—all this time—were full of pee?

ANGEL ON SHOULDER

Wait.

Don't jump to conclusions.

DEVIL ON SHOULDER

Whaddya mean, *jump*? You literally just did the experiment!

ANGEL

Yes, but there are lots of other things that come off our bodies besides pee. Like . . . sunscreen!

DEVIL

Hm, I guess you're right. Spit comes out of your body, too.

ANGEL

Yeah! That would be less gross.

DEVIL

And snot.

ANGEL

Ugh. Really?

DEVIL

And shit.

ANGEL

You're disgusting.

The angel and the devil bring up four interesting possibilities: sunscreen, spit, snot, and poo. I tested two. Snot produced a mild, sort of half-assed Pool Smell. Spit produced—to my nose—the canonical Pool Smell, exactly like what you smell walking into a gym's indoor pool.

So we can now take our original statement—"Chlorine is associated with the Pool Smell"—and modify it slightly. Something like this might be appropriate: "Chlorine, when mixed with human pee, snot, or spit, produces a smell remarkably similar to the smell of an indoor public pool." As is often the case in science, doing an experiment can raise more questions than it answers. How much of a pool's smell is caused by pee, and how much is caused by snot or spit? Is the smell produced in my small-scale experiment chemically identical to the one in public pools? Do people *really* pee in public pools so frequently that every single one smells like the Pool Smell? (I think we all know the answer to this one.) You could spend your whole career just studying the chemistry of public pools.[*]

Pools and pee are the crux of pothole #7: study design. In other words, the type of study you run can limit what you can conclude about whether something is causal. Going around sniffing public pools is an observational study. It can tell you that there's an association between chlorine and the Pool Smell, but it can't tell you that chlorine *causes* the Pool Smell. Peeing in one bucket and not

....
[*] And, of course, people have.

peeing in another, then comparing the smells, is a controlled trial. It can tell you that pee and chlorine cause the Pool Smell. But it has its own limitations. (More on those in a few pages.) And convincingly figuring out the chemistry of *how* chlorine and pee (or other bodily fluids) react to form the Pool Smell is another layer in the evidence cake, just like figuring out the chemistry of how smoking causes lung cancer is another brick in the Bridge of Truth.

So the next time you see a headline like

BLUEBERRIES LINKED TO REDUCED RISK OF DEATH

remember all the potholes on the road to a legit and causal association:

> **Pothole #1: fraud**
>
> **Pothole #2: basic mathematical mess-ups**
>
> **Pothole #3: Random Chance**
>
> **Pothole #4: procedural problems**
>
> **Pothole #5: statistical skulduggery, including p-hacking**
>
> **Pothole #6: confounded associations**
>
> **Pothole #7: study design (observational vs. randomized controlled trial)**

There are other potholes, but these are the main ones to remember—

Wait a second.

Why is it *your* job to remember all this shit? Shouldn't it be . . . literally anyone else's? Picking apart studies like this—finding all the potholes and trying to decide if they affect the final result—is really hard work. Wouldn't it be great if there were a group of scientists who developed a systematic way to rate studies and help you decide whether or not you should stop eating Cheetos?

Luckily, there kinda is.

Pretty recently, a group of scientists developed a systematic way to look at a bunch of evidence and decide how good (or bad) it is. Thus was the GRADE System born. The levels of the GRADE System—which you can think of as the letter grades you got in school—look something like this:

NO FRIGGIN CLUE

THAT TIME YOU WORE RED UNDERWEAR AND SCORED THREE GOALS — BASICALLY NO EVIDENCE ("MIGHT AS WELL BE MAGIC")

THE MASSAGE/OBESITY STUDY FROM CHAPTER 7 — BAD SCIENCE ("LITERALLY INCREDIBLE; THERE MUST BE SOME MISTAKE")

A FEW WELL-DONE OBSERVATIONAL STUDIES; NO RANDOMIZED CONTROLLED TRIALS — LOW-QUALITY EVIDENCE ("HUH, INTERESTING; WORTH STUDYING MORE")

A FEW WELL-DONE RANDOMIZED CONTROLLED TRIALS — MEDIUM-QUALITY EVIDENCE ("THIS ACTUALLY MIGHT BE TRUE")

MANY LARGE, WELL-DONE RANDOMIZED CONTROLLED TRIALS WITH CONSISTENT RESULTS, IN THE SPECIFIC POPULATION YOU'RE INTERESTED IN, OR, MANY LARGE, WELL-DONE OBSERVATIONAL STUDIES WITH CONSISTENT HUGE EFFECTS AND A DOSE-RESPONSE RELATIONSHIP (E.G., SMOKING AND LUNG CANCER) — HIGH-QUALITY EVIDENCE ("WE'RE ALMOST POSITIVE THIS IS TRUE")

TRUTH

As I write this, most of the evidence against ultra-processed foods is observational, and the risks are not huge, which makes it low-quality. Low-quality evidence is the kind of evidence that you look at and go, "Huh, that's interesting. Maybe we should do a randomized controlled trial to test if this association is legit and causal." It's *not* the kind of evidence that you look at and go, "We've conclusively concluded with 100 percent conclusivity that this association is legit and causal. Alert the media!" To be clear, all I'm saying here is that we should be cautious about the evidence on ultra-processed food. I'm *definitely not* saying that eating processed food is *good* for you.

Ultra-processed food could turn out to be an important cause of obesity and diabetes; but it could also turn out to be mostly a passenger variable, tapping on the death accelerator a little, but not the major driving force. We don't know yet. Research dollars will continue to be spent on studying ultra-processed foods, so the evidence will eventually get better. Who knows: one day it might even be good enough to start imprisoning bags of Cheetos. But that'll take years.

About forty-two seconds after I typed that sentence, I got an e-mail from Kevin Hall, a metabolism researcher at the National Institutes of Health who had done some studies with contestants from *The Biggest Loser*. I had reached out to him a few weeks before to ask some general questions about processed food, obesity, and metabolism. Kevin replied: "The processed-food manuscript is currently under peer review, so I can't discuss that study yet . . ."

Whaaaa???

I had no clue, but it turned out that his group was about to publish the first randomized controlled trial using the NOVA food classification scheme we talked about in the very first chapter.

Journalistic instincts be damned; all you need is a little Random Chance.

Hall's study was the first randomized controlled trial to test whether ultra-processed foods make you eat more calories and gain more weight than unprocessed foods. Let me just say that studies like these are hard and expensive. Why? Well, remember at the very beginning of the book when we talked about locking two groups of people up on separate deserted islands, feeding them different diets, and then seeing what happens over decades? Hall and his team basically did exactly that, except for a period of twenty-eight days instead of years, and at the NIH hospital in Bethesda, Maryland, instead of on two deserted islands. Despite those differences, this study was still a major pain in the ass. Hall had to find people who were willing to:

- live in a hospital for a month, never leaving it once
- eat only the food they were given within sixty minutes and have someone take away whatever was left on their plate to weigh it
- get on a scale every morning at 6:00 A.M. and have a nurse record their weight
- get x-rayed every week
- get MRI'd every two weeks
- pee in a cup daily
- get locked in an airtight chamber for about twenty-four hours every week to measure their energy expenditure
- have their blood drawn three times in four weeks
- wear an accelerometer twenty-four hours per day to measure physical activity

- **do twenty minutes of stationary biking thrice daily**

Finding twenty healthy human volunteers to sign up for—and complete—Hall's study was basically a miracle. Seriously, hats off to these folks—taking one for science!

So how did the study actually work? It's pretty simple: The twenty volunteers were randomly split up into two groups of ten each and started on either an ultra-processed or unprocessed diet. Both diets had roughly the same number of calories, proteins, carbs, and fat. The major difference was whether calories came from ultra-processed or unprocessed food. (There were other differences, but we'll get into that later.) After two weeks, everyone switched diets: everyone on the ultra-processed diet switched over to the unprocessed diet and vice versa. In both diets, volunteers were given double the number of calories needed to maintain their weight. Why double? Because Hall and his team were trying to figure out whether people would eat more food if it was ultra-processed, and the only way to do that is to give people essentially unlimited quantities of food and let them eat as much as they want.

I have to say, never did the *Sorcerer's Stone/Failure to Bone* thing hit home harder than when I was reading menus of the two different diets. For example, on day five, dinner for the unprocessed group was grilled beef tender roast, barley with olive oil and garlic, steamed broccoli, a green salad with vinaigrette, and apple slices. On day seven, dinner for the ultra-processed group was PB&J sandwiches on white bread, Cheetos (baked), graham crackers, chocolate pudding, and 2 percent milk. On some days the ultra-processed diet didn't seem quite so bad—breakfast on day one was Honey Nut Cheerios, a blueberry muffin, margarine, and whole milk—but overall, it was definitely a *Failure to Bone* situation.

I bet you can guess the results: people on the ultra-processed diet ate more calories (about 500 more) and they also put on about 0.9 kilograms over the course of the trial. People on the unprocessed diet *lost* about 0.9 kilograms. And remember, this wasn't an observational study. It was a (cue angelic choir) randomized controlled trial.

Seems pretty solid, right?

Yes. But of course there's no such thing as a perfect experiment. So let's put our Asshole Hat back on.

To test whether ultra-processing food *causes* people to eat more and put on weight, you have to—as scientists say—"isolate the variable of interest." This just means you have to make sure that the *only* difference between the two diets is whether the food is ultra-processed. Why? Imagine if, in my experiment to figure out whether pee causes The Pool Smell, I had filled the beakers as follows:

Beaker #1: distilled water + chlorine

Beaker #2: garden hose water + chlorine + pee

Beaker #2 would have smelled Of Pool . . . but I would not have been able to conclude that pee was the culprit. Why? Because there could have been something in the garden hose water that was causing The Smell, or reacting with chlorine to cause The Smell.

With a simple experiment like that, it's relatively easy to isolate the variable of interest. But in a study involving an entire diet, it's *much* harder. Even though Hall and colleagues tried their darnedest to make sure the two diets were as similar as possible on all the variables, some—like "calories per gram"—were stubbornly unmatchable. "Calories per gram" is also known as energy density, and foods can differ *wildly* on this particular variable. For example, a slice of Peppermint Bark Cheesecake at The Cheesecake Factory holds a whopping 1,500 calories, whereas a similar mass of whole milk has about 130 calories. As we know from chapter one, ultra-processed foods are very energy dense. And as it turns out, energy density can make you eat more totally *independently* of food processing. Think of Guy Fieri cooking you a meal

His "meal" would be extremely energy dense, but totally unprocessed. Now think of Giada De Laurentiis cooking you a meal. It would also be unprocessed, but much less energy dense. Which would you eat more of?

Yeah, Guy's, and you'd hate yourself the whole time.

Same idea in Kevin Hall's study. The ultra-processed diet was much more energy dense.[*] So the weight gain could have been caused—at least in part—by the difference in energy density, and not *just* the difference in processing.

If you happen to be screaming at the page that nobody in their right mind would choose Fieri's Fajita Fiesta over Giada's Cannellini Spezzatino, that's because there's another variable we haven't considered yet: taste.

Maybe, as one tweeter pointed out, "the study simply found people like eating tasty quesadillas more than boring salad"; in other words, maybe people just like *Failure to Bone* more than *Sorcerer's Stone* because they prefer porn to art. It's not a crazy thought, but the twenty people in the study rated the two diets as roughly the same in terms of "pleasantness." You could argue this means taste wasn't a factor, but Dennis Bier, the former editor in chief of the *American Journal of Clinical Nutrition*, disagrees. He thinks the very fact that people ate 500 more calories on the ultra-processed diet is a very plausible indicator that the ultra-processed diet *did* taste better.[†]

So, if the study really was intended to test how ultra-processing

....

* Excluding drinks, the ultra-processed diet was almost twice as energy dense as the unprocessed diet. To try and make up the difference, Hall and his team dissolved a whole bunch of fiber into diet lemonade and added that to the ultra-processed diet. But that's not the same as matching the solid food portions of both diets.

† How is this possible, given that people rated the diets the same? Maybe the participants artificially inflated the taste of unprocessed food because it felt "good" or "natural." Or maybe they inflated the taste because they knew they were being studied.

food *alone* influenced weight gain, the diets should have been more closely matched in terms of energy density and a few other variables we haven't considered. But there are other reasons that we might not want to take this finding to the bank just yet.

This study was small (twenty people) and relatively short (twenty-eight days), especially when you compare it to a lifetime of eating. Also, Hall couldn't blind people to the food they were eating. That would have been impossible. He did try to blind volunteers to certain aspects of the study; for example, they were told this was *not* a weight-loss study and blinded to their weight measurements. But the participants must have intuitively known they were in a study designed to test whether processed food was bad for you, and they might have had preexisting beliefs about ultra-processed food that could have affected the results.

Also, the environment of this trial was nothing like real life. In addition to living in a hospital and having everything (except for their shit) measured, volunteers were regularly asked things like "How hungry do you feel right now?" and "How much do you want to eat right now?" They were also asked to rate the food they were eating, sometimes literally *while* they were eating it. What's the problem with this? It's not so much that it would affect the difference between the two groups; it's more that it might affect whether a similar experiment would work *outside* of this particular setting. In other words, if you went home and replicated the unprocessed diet exactly, you would not be replicating the *full* experiment: you wouldn't be living in a hospital setting; you wouldn't be constantly poked, prodded, and weighed; and you wouldn't be thinking about food nearly as much. All of those things could very well affect your behavior, which could affect whether you eat fewer calories and lose weight. But there's not much Hall could do about that. Imprisoning volunteers in a hospital—and thus drastically changing their natural environment—was the only effective way of making absolutely, positively sure he knew what they ate.

Another possible issue: the volunteers themselves. On average, they were solidly within what the World Health Organization considers "overweight," with a BMI of 27. They were also pretty young (average age: thirty-one). And they were willing to participate in a monthlong, complicated clinical trial. Again, that probably wouldn't affect the difference between the two groups, but it might mean that the results simply don't apply to you. If you're, say, seventy-five years old, have a BMI of 22, and have no inclination to allow scientists to study you, your body might be sufficiently different from the participants' that the results of the study wouldn't transfer over from them to you.

Both of the issues above are general issues with all randomized controlled trials, not specific things that Hall did wrong. In fact, the most common critique of randomized controlled trials is that the setup of the study—which includes the environment and the people being tested—almost never applies *exactly* to your situation. So you can't necessarily generalize the results to the set of people you're interested in.

Okay, let's take the Asshole Hat off and give Hall some major props where major props are due.

First, this study was thoroughly and thoughtfully preregistered. Hall called his shots and measured exactly what he said he was going to measure. Also, he chose to make all his raw data freely available, which means that if anyone wants to go through and check every single calculation—or do additional ones—they can, without asking permission. Both of these choices make me confident that this study is not a pile of p-hacked bullshit. And even though some of the variables were different between the two diets, many were very similar. For example, the diets were almost identical in the proportion of calories that came from carbs, protein, and fat. So you could cross some variables off the list of potential culprits—and that's very useful.

So, does this trial change how I see the evidence against ultra-processed foods?

Yes, a little.

This trial shows that ultra-processed food *caused* a group of overweight young people to put on weight. But—stick with me here—we cannot conclude that processing was the reason that ultra-processed foods caused the weight gain.

At first glance, this makes zero sense. How can we know that ultra-processed foods did something, but not know whether they did it because they were processed? It comes down to the fact that ultra-processed foods are a whole bunch of variables tied together in one convenient package: energy density (high), volume (low), flavor (delicious), location of production (factory), salt level (high), and so on. Because some of these variables remained stubbornly unmatched in Hall's study, it's impossible to tell which of them was responsible for the weight gain. Was it the processing? Maybe. But maybe not. It could have been the energy density. Or the type of fiber. Or the taste. And so on.

You can think of it like the distinction between knowing *that* ultra-processed food causes weight gain vs. knowing *why* it does. We always want to know *why*, but sometimes you have to start with the *that*. Hall's study is a necessary and important first step down the path. More experiments will follow.

Also, we shouldn't forget that the trial was short and small and was carried out in a specific population and in a very structured environment; so the results may not apply as broadly as we instinctively believe that they do.

Overall, I'd say this is a decent starting brick in the just-starting-to-be-constructed Bridge of Truth on ultra-processed foods. But I also think we need more bricks and more mortar before we can truthfully say we know THE ANSWER with certainty.

YOU'RE LATE FOR A VERY IMPORTANT DATE

This chapter is about memory, failed children,
rabbit holes, warts, and death.

We've covered seven potholes to a *legit* and *causal* association. Now we wade into contentious waters: Do these potholes affect nutritional epidemiology? To find out, I called biostatistician Betsy Ogburn, who told me I was thinking about it wrong. "If you asked most nutritional epidemiologists to describe the weaknesses of their studies, they would list all of these things." In other words, everyone agrees that the potholes are there.

But, she continued,

I think it's really hard to actually internalize the fact that any one of these can undermine what looks like very strong evidence, especially if a researcher has poured their blood, sweat, and tears into a study.

Ogburn was right. The nutritional epidemiologists I spoke to acknowledged that the potholes exist; they all agreed that you can't barrel down the street of science and simply pray that you don't hit one. But they vehemently disagreed on two points: First, did the car even hit a pothole? And second, is the car still drivable?

It seems weird that two scientists can stand at the beginning of a road, clearly see potholes in front of them, agree that the potholes are there and dangerous, then get in the car, drive to the other side, and yell at each other about whether they hit anything and if the car is totaled or not.

To understand why this is so hotly debated, let's travel back to the late nineties and early aughts. In 2005, a Greek epidemiologist named John Ioannidis published a paper with the completely benign and nonprovocative title "Why Most Published Research Findings Are False." If you were following science news back then, you read about this paper. And whether you agreed or disagreed with the title, it indisputably made a big splash in the scientific community. It helped spur the launch of several reproducibility projects in psychology and basic cancer research (in which scientists redo published experiments to see if they can replicate the results). Ioannidis moved to Stanford and turned his attention to nutritional epidemiology, telling a Canadian reporter that "it should just go to the waste bin" and a Vox reporter that it was "a field that's grown old and died. At some point, we need to bury the corpse and move on . . ."

Them's fightin' words.

Nutritional epidemiologists are fighting back, though they don't quite have Ioannidis's flair for the dramatic. Walter Willett, a nutritional epidemiologist at Harvard, responded with "You really seriously misrepresent how nutritional epidemiology is conducted," which is kind of like responding to "Everything you stand for is a rotting pile of shit" with "We have different opinions on the subject." The gulf between Ioannidis and Willett—which I will henceforth call the nutritional epidemiology war—is by far the biggest and deepest rabbit hole I found while researching this book. If your idea of fun includes horrifically confusing statistics,

Harvard-Stanford rivalry,[*] and/or questioning your sanity, you will *love* this particular rabbit hole. My idea of fun includes none of those things, so, rather than diving in, I will gently guide you around the periphery of the hole. We will lean forward and peer into the yawning blackness and interrogate the characters within, but We Shall Not Pass the event horizon.

One of Ioannidis's arguments against nutritional epidemiology is what we talked about in chapter seven: the more hypotheses you test, the more likely at least one is to turn out "statistically significant" by Random Chance. But Ioannidis argues that this is *especially* true of food and disease, because there are basically infinity hypotheses to test:

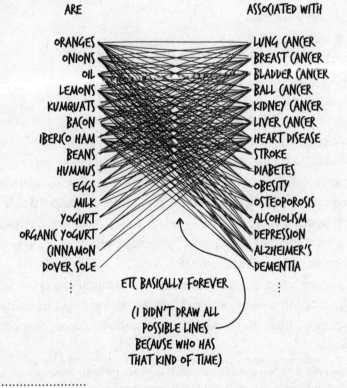

....

* In 2010, Harvard made Ioannidis an adjunct professor of epidemiology, clearly staking out a position—on both sides of the war. Classic Harvard.

Each one of the lines is a potential experiment. If you had, let's say, 300 different foods and 800 diseases, you'd have 240,000 potential experiments. Even if only 5 percent of those show a statistically significant association by Random Chance, that's still *12,000* experiments showing an association between, say, kumquats and anal abscesses.* Ioannidis also contends—and I agree—that results showing a link between food and disease are much more likely to get published and get media coverage, so you're much more likely to read

DOVER SOLE ASSOCIATED WITH 23% HIGHER RISK OF TESTICULAR CANCER

in the news than

BAY LEAVES NOT ASSOCIATED WITH ANYTHING IN PARTICULAR.†

Willett and company's counterargument is: We don't just blindly test every hypothesis available like a robot. We use our knowledge of biochemistry, animal experiments, and metabolic experiments to narrow down the hypothesis list to the most plausible ones. Also, we've recently moved from testing single foods to testing dietary patterns (like the Mediterranean diet), which reduces the total number of available hypotheses and more closely reflects how people eat in real life.

Another of Ioannidis's arguments is that nutritional epidemiology is based on observational research rather than randomized controlled trials. Remember, in observational studies, you don't

....

* It's a little more complicated than this. Some of the variables are not independent, so the real numbers are lower—but the basic idea is the same.
† Just to be clear, I made both of these headlines up.

actually try to change people's behavior. In the best-designed ones, you recruit a group of people, follow them over a period of years until some of them get cancer, heart disease, or whatever outcome you're interested in studying, and then compare the people who got cancer (or heart disease) with the people who didn't. Did they smoke more? Exercise less? Eat fewer mole rats?

We'll get to Ioannidis's issues with observational studies, but first let's give them their due: Willett argues—and I agree—that observational studies have had some success, the most famous and best example being smoking. Early evidence against smoking was mostly observational. Remember, the number of randomized controlled trials that the surgeon general cited in the 1964 report was . . . zero. Granted, that's not technically *nutritional* epidemiology, since you don't eat cigarettes, but nutritional epidemiology has also had its observational trial high points. Team Willett cite a few examples, most recently trans fat. They also say they don't *just* rely on observational studies—they use randomized controlled trials, too—but as they rightly point out, observational studies are cheaper, and sometimes randomized controlled trials aren't ethical or practical for what you want to study.

Okay, so what's Ioannidis's beef with observational studies? First beef: they're observational—the equivalent of going around sniffing public pools, not peeing in a cup—so even if they *could* reveal legit associations between a food and a disease, they *cannot* reveal if a food is actually *causing* the disease.

Second beef: memory. This is a long one, so buckle up.

Consider this tale from Indiana, which begins innocently enough: "Last week, a Marion County Sheriff's Office employee purchased a McChicken sandwich from the McDonald's location at 3828 W Morris Street, Indianapolis." Then he put his sandwich in the fridge and went about his business. Seven hours later, he returned to find *A BITE OF HIS SANDWICH WAS MISSING*!

Dun
    Dun
        Dun

He immediately concluded that "a McDonald's restaurant employee had tampered with his food because he is a law enforcement officer." So he went back to McDonald's and complained. According to the *Washington Post*, which covered the story, both McDonald's and the Marion County Sheriff's Office launched *full investigations* into The Case of the Missing Bite.

What was the solution to this particular mystery?

"The employee took a bite out of the sandwich upon starting his shift at the Marion County Jail, then placed it in the refrigerator in a break room. He returned nearly seven hours later having forgotten that he had previously bitten the sandwich."

In case you're wondering: yes, this really happened. I'm quoting from a statement summarizing the events, issued by the Marion County Sheriff's Office.

What's the point of this story? It's this: you suck at remembering what you ate.

Remember that the whole point of nutritional epidemiology is to try and figure out whether what you ate causes disease. So if nutritional epidemiologists can't reliably figure out what you ate, there's no way they can reliably associate foods or diets with diseases. It's just not possible. For this reason, a decent-sized chunk of nutritional epidemiology (and the criticism thereof) is focused on the critical question of how much your memory sucks (or doesn't). Now, Willett would say that The Case of the Missing Bite proves nothing at all. And he'd be right: a story is not science. So let's talk science.

Ideally, there would be a simple, easy, cheap, and accurate way to figure out how good or bad people's memory is when it comes to food. Unfortunately, there isn't. In fact, evaluating memory about *anything* is hard. And memory is only a part of the larger question: how reliable is what people say? Suppose you ask someone how often they go to the gym and they say "three times a week." Should you believe them?

There are a few ways to poke at this question. Let me start by introducing you to my friend NHANES, or the National Health and Nutrition Examination Survey. NHANES is run by the Centers for Disease Control and Prevention, affectionally known as the CDC. Every year, they:

1. **pick a representative sample of 5,000 Americans.**

2. **give them the most comprehensive exam they've ever had or will ever have.**

Medical history. Family medical history. Physical exam. Dental exam. Blood tests. A hearing test. Physical activity monitoring. A pregnancy test. Diet questions. NHANES asks you how much money you make; what color your skin is; whether and how often you smoke, exercise, have sex (with or without birth control, condoms, dental dams, etc.), and use drugs; and pretty much every other question it is possible to ask someone without getting thrown out of their house. This line of questioning continues for almost an entire day or until the participant dies of exhaustion.

I'm being facetious here; NHANES is actually a Herculean effort on behalf of both the participants and the staff. It costs over $100 million to collect this data from just 5,000 people. The result is a mind-bending amount of information: Imagine if, when you went to the doctor, instead of shooing you away after twelve minutes, they invaded your privacy with a whole day of questions and then ran every single test LabCorp and Quest Diagnostics offer.

NHANES also does two simple but ingenious things:

1. They *measure* your height and weight.

2. They *ask* you what your height and weight are.

And that makes it possible to compare what you *say* your height and weight are to what your height and weight *actually* are. This is a simple way to test whether you can trust what someone tells you about their height and weight, which is exactly what two scientists did back in 2009. They downloaded data from three rounds of NHANES—about 16,800 people—and compared what people said to their actual measurements.

Results?

On average, men said they were half an inch taller (and a third of a pound heavier) than they really were; women said they were a quarter inch taller (and three pounds lighter) than they really were.

If you've spent any time on Tinder, these results ring true. In fact, at least for me, they're comically accurate: the only time I was ever asked how tall I was on Tinder, I added half an inch.

In the study, there wasn't a single group of people—young, old, rich, poor, underweight, overweight—who said they were shorter than they actually were. Everyone thinks they're taller than they are. But when it comes to weight, the story is a little more interesting. Men almost universally said they were heavier than they really were—except men defined as obese by the CDC (having a BMI over 30), who, on average, said they were three pounds lighter. Women almost universally said they were lighter than they really were (except underweight women, who said they were heavier).

Who do you think was the *worst* group of people at accurately estimating their weight?

It wasn't who you think.

It was underweight men. Specifically, men defined as underweight by the CDC said they were a full eight pounds *heavier* than they really were. The 2009 study wasn't an outlier—there are lots of studies showing that what people *tell you* about themselves can be very different than what you *measure*.

When it comes to height and weight, that doesn't seem like a big deal. What's half an inch here and a few pounds there? To some extent, I agree. But height and weight are simple, so shouldn't the error be smaller? And that makes you wonder: If people get their own height and weight wrong, how reliable is what they tell you about their diet?

Every scientist I talked to—including Ioannidis and Willett—agreed that food is complicated . . . much more complicated than weight. You probably eat hundreds, maybe even a thousand, different foods over the course of a year, in wildly different quantities. You eat differently as the seasons change. You cook at home; you eat out; you snack; occasionally, you fast or binge. Over the course of years, your diet can change dramatically.

There are so many more ways to be wrong when it comes to food.

But before we get to that, let's talk about how food is measured in the first place. Most of the time, unlike in Kevin Hall's randomized controlled trial, food is measured by a "memory-based method," which is exactly what it sounds like: you tell someone what you ate. But there are many different flavors of memory-based methods. For example, many studies use what are called "twenty-four-hour recalls" which are also exactly what they sound like: you tell someone what you ate in the past twenty-four hours. NHANES uses two twenty-four-hour recalls, each one administered in five "passes." Basically, you tell them what you ate over the past twenty-four hours . . . five times. Why five times? Because your memory gets better each time. You remember additional things you ate—that Klondike bar for Lunchdessert—on the fifth pass that you didn't on the second.

Other studies use food frequency questionnaires. Every research group has their own flavor of food frequency questionnaire, but the most detailed ones usually ask:

how often

on average

you ate a specific quantity

of a category of food

over the past year.

For example: How often, on average, did you eat six ounces (or one serving) of french fries, over the past year?

- **Never.**
- **Less than once per month.**
- **1–3 times per month.**
- **Once per week.**
- **2–4 times per week.**
- **5–6 times per week.**
- **1 or more servings per day.**

Before we get to memory, there's an even more basic issue: understanding the question.

Why does it feel like you need a day planner just to figure out the difference between "five to six times per week" and "one or more servings per day"? And speaking of servings, is it "six ounces" or "one serving"? Those are the same only when . . .

they're the same. Which is to say, they're often not. In fact, none of the sizes of fries at McDonald's are exactly six ounces (a Large is about 5.3 ounces) . . . unless, of course, they mean six ounces *by volume*, in which case . . . how tightly do you smash the fries into the ¾-cup measure? (Let's assume not tightly.)

Incidentally, are we talking McDonald's fries, classy restaurant fries, or fries you make at home? Remember from chapter one that a chemist would say they're all the same; Carlos Monteiro would probably beg to differ. And then there's "on average," which is usually explained like this: "Please try to average your seasonal use of foods over the entire year. For example, if a food such as cantaloupe is eaten four times a week during the three months that it is in season, then the *average* total use would be once per week over the year."

Quick math translation, for those of you as confused as I was:

$$4 \; \frac{\text{TIMES}}{\text{WEEK}} \; \times \; 4.33 \; \frac{\text{WEEKS}}{\text{MONTH}} \; \times \; 3 \text{ MONTHS} = 52 \text{ TIMES (THAT ONE QUARTER OF A MELON WAS EATEN DURING THE SUMMER)}$$

$$\text{So...} \quad \frac{52 \text{ TIMES IN THE SUMMER}}{52 \text{ WEEKS IN THE YEAR}} = \text{ONCE PER WEEK (ON AVERAGE)} \quad \text{PHEW!}$$

Sometimes, the way food is categorized seems to make no sense. For example, there's one question about "tortillas: corn or flour . . ." and another question right below it about "potato chips or corn/tortilla chips . . ." I understand the desire to split tortillas from tortilla chips, but lumping in tortilla chips with potato chips? That's an entirely different species of plant.

Some of the questions, if taken literally, would produce wildly wrong answers. For example: How often, on average, did you eat two slices of pizza over the past year?

How often did I eat *exactly two* slices of pizza?

I've never eaten *only* two slices of pizza in my entire life.[*]

The bottom line, says epidemiologist Katherine Flegal, is that "these are very difficult questions to answer; they're not cognitively friendly. People don't naturally think that way."

And there's the issue of memory. Flegal continues:

> People know foods they never eat. No problem. "I hate kale, I never eat it." You're done. And they know the foods they eat every single day. "I have this for breakfast every single morning of my life." What they don't know is everything in between, which is most food.

Last but not least, some people—*gasp!*—lie about what they ate.

It might seem crazy that a decent chunk of nutritional epidemiology is based on these types of dietary surveys, but let's consider some of the counterarguments for a moment. Briefly, they are:

1. **Nobody expects memory-based methods to be absolutely perfect.**

2. **They don't *need* to be perfect; they only need to be good enough.[†]**

3. **Dietary measurement error is "nondifferential."**

You don't have to worry about what that last one means; the point is that memory-based methods tend to *under*estimate the relative risk. So, for example, remember that study that found an

....

[*] I'm being facetious; I understand the point of the question. Still, it's weird.

[†] I'm skipping over counterargument 2(b), which is "We've validated memory-based methods." The debate over whether food frequency questionnaires and other memory-based methods have actually been validated is a thorny morass that I won't get into here. From what I can tell, the argument boils down what counts as "good enough."

association between ultra-processed foods and a 14 percent increase in risk of death? If we accept that the only type of error was dietary measurement error, then the real risk is probably higher than 14 percent. Exactly how high depends on how bad the measurement error was and how well the researchers mathematically adjusted for it.

So what to make of all this?

Intuitively, memory-based methods seem . . . flimsy. But their proponents say they're good enough to detect associations between foods and disease. Plus, they say, memory-based methods are all we got. And they're right: as of right now, and to the best of my knowledge, there is no other way to cheaply estimate what people eat over the course of decades. Then again, critics of memory-based methods also make a good point: if a particular method isn't good enough, you shouldn't use it at all—even if it's all you've got.

There is perhaps no more debated point in nutritional epidemiology than this: how reliable are these questionnaires? Willett and company say they're reliable enough to make public health pronouncements like "After considering all available evidence, including observational studies, animal experiments, and controlled trials of intermediate endpoints, we can reasonably conclude that bacon gives you butt cancer"; Ioannidis and company say they're essentially worthless. So . . . yeah, the parties are pretty far apart on this one.

Ioannidis's third beef with nutritional epidemiology sounds kind of boring. It's this: most nutritional variables are densely associated with one another.

What does this mean, and who cares?

It basically means that if you eat an apple a day, you are

unlikely to also drink a Sonic milkshake every day. Or if you make $80,000 a year, you're more likely to eat a lot of avocado toasts and drink soy lattes between hot yoga classes. Or if you work out regularly, you're likely to eat more chicken than T-bone steak. Basically, the idea is that nutritional and lifestyle variables—like food intake, physical activity, how much money you make, whether you smoke, how long you live, and so on—are much more associated with one another than variables in other scientific fields. This, by itself, is not an earth-shattering insight. Of course apple consumption is closely related to, say, carrot consumption: if someone tries to lead a healthy lifestyle, they're more likely to eat both apples and carrots. Association (potentially) explained. Ioannidis's argument is that *so many* of these variables are associated with one another that it essentially makes the whole exercise of nutritional epidemiology useless.

Why?

In Ioannidis's view, finding a statistically significant association is kind of like finding out that Taye Diggs follows you on Twitter: amazing at first, but pretty meaningless when you find out that Taye Diggs follows pretty much everyone on Twitter.

Let's do a thought experiment. Say you do an observational study and find that eating an apple a day is associated with a 22 percent reduced risk of death. Seems fairly clear:

APPLE •————————• 22% LOWER RISK OF DEATH

But if you had kept looking, you would also have found that eating an apple a day is also associated with eating fruitcakes, carrots, ginger tea, and exercising. Because eating apples is associated with a reduced risk of death, all those other things are *also* associated with a reduced risk of death via their association with apples. Now our (hypothetical) picture is a little bit more convoluted:

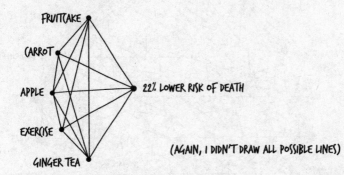

FRUITCAKE
CARROT
APPLE
EXERCISE
GINGER TEA

22% LOWER RISK OF DEATH

(AGAIN, I DIDN'T DRAW ALL POSSIBLE LINES)

And this is just my third-grade-level cartoon. On the page opposite is what an actual "association globe" might look like for just nineteen common nutritional measurements—things like how much fat, protein, carbs, fiber, alcohol, and veggies you eat, as well as blood test results like vitamin, mineral, and cholesterol levels. This globe, unlike my cartoon, is based on real data.

If it looks like everything is pretty much associated with everything else, that's because . . . it is. The question then becomes: Which variable is driving the outcome (could be cancer, heart disease, death, or anything else you're studying) and which ones are along for the ride? More realistically, some variables might be flooring the accelerator, some might be gently pressing it, some might be gently tapping on the brakes, and others might be sitting in the back, doing nothing but *COMPLAINING AND TEXTING* GODDAMN IT EVAN GET OFF YOUR PHONE.

This is another nub of the disagreement between Willett and Ioannidis. Willett would tell you that the mathematical methods used to adjust for all these variables are robust and, wielded by the proper researchers, can produce results we can all be confident in. Ioannidis would tell you that that might be true when

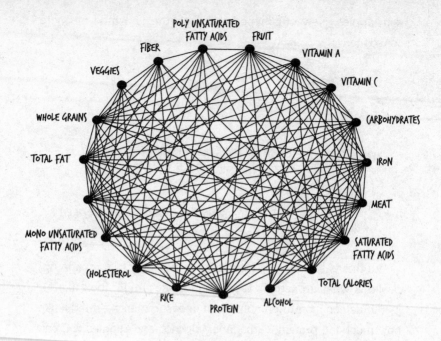

AND I SURE AS SHIT DIDN'T DRAW ALL THE LINES HERE EITHER

you're talking about smoking-sized risks, in the range of 1,000 percent, but it's not remotely true for much smaller risks, like the 14 percent increase in risk of death associated with ultra-processed food. I have to say, I agree with Ioannidis on this one.

Ioannidis's contention is that figuring out which variables are really in the driver's seat is almost impossible without randomized controlled trials.

Willet's contention is that well-performed observational studies can adjust for confounding variables well enough to make public health pronouncements.

We've now completed our tour of the periphery of the war. I've necessarily simplified the arguments and the cast of characters. Behind each person—Ioannidis and Willett—there are others. Behind each argument there is a rebuttal. Behind those rebuttals are more rebuttals. The rabbit hole goes down as deep as you are willing to dive, and at the bottom I suspect there is only pride and fear.

———

Remember the Asshole Hat we first donned back in chapter seven? Wearing it is undoubtedly fun, but at some point you have to switch out the Asshole Hat for a hard hat and start trying to fix problems instead of just pointing them out. Scientists are a fixer-upper bunch, so there is no shortage of potential solutions being proposed.

Let's go on a (very brief) tour.

The most dramatic—and hotly debated—solution is to cut down on the number of observational studies and instead spend the money on large, randomized controlled trials. You can guess who's at the vanguard of this position and who's vehemently against it. Perhaps cynically, I suspect this disagreement will only be resolved by who retires first . . . or who's in charge of doling out NIH's money.

What about p-hacking?

One solution is to do what Kevin Hall did. Call your shots. Preregister your study, especially your data analysis plan. Another solution is to show your work. Not just "Here I will briefly summarize the main points"; instead, show *every single step*. Brian Nosek made this point repeatedly in our interview. He even emphasized it over preregistration:

> All I want is for people to show how they got to the claims
> that they made, whatever way they did it. Totally
> exploratory, totally strong preregistration, multiverse
> analysis when you analyze it tons of ways—doesn't matter.
> Just show the work from start of the idea to when you drew
> that conclusion.

Part of showing every detail of how you did something includes releasing the raw data—anonymized, of course—and the code you used to analyze it. This is anathema to some researchers, but others are embracing it. For example, if you wanted to download and reanalyze every ounce of raw data from Kevin Hall's processed food study, you could.* If Walter Willett wanted to try and poke holes in the study, he could download the raw data. And if John Ioannidis wanted to poke holes in Willett's hole-poking, he could *also* download the raw data and do that. And if a random third party wanted to plug some holes and poke others, they could do that, too, using the identical data set that everyone else already has. This kind of thing is already happening, and I think it's great. I'm 100 percent on board with full and open data sharing.

Another solution to the p-hacking problem is to set up something called a "specification curve." To understand how this works, first consider a recipe for chocolate chip cookies. Even with a fixed and immutable list of ingredients, you still have a lot of flexibility in terms of how you combine those ingredients to make cookies. You could follow the recipe exactly, but you could also tweak lots of things; for example, you could bump the oven temperature up 15 degrees, wait until the butter hits room temperature before creaming it or not, freeze the dough for twenty minutes before it goes in the oven, or spread the pinch of salt out over each individual cookie instead of adding it to the batter. The possibilities are infinite. The same thing is true of a study: even with identical data (ingredients), there are a lot of different ways to analyze that data, and those differences can produce very different outcomes (cookies). That's part of what makes p-hacking possible.

Most of the time, researchers simply choose whatever data analysis method they feel is best and go with that. Problem is, not

....

* It's at https://osf.io/rx6vm/.

everyone agrees on what's best. This is where the specification curve comes in. Its approach is to do *ALL THE RECIPES!* So instead of making one batch of cookies, you make hundreds, systematically tweaking every possible variable and seeing how that affects the taste. Ditto for science: you make a computer crunch the numbers every possible way and see how that affects the results. If the results are broadly the same no matter how you crunch the numbers, you can be pretty sure you're onto something real. But if tweaking the data analysis recipe sends your results all over the damn place, maybe the effect isn't quite as real as you first thought.

Some solutions have almost nothing to do with science and more to do with . . . common sense.

If your reaction to the Willett/Ioannidis disagreement was *How is it possible that two wildly intelligent people can disagree so vehemently on something that seems so . . . mathematical!?* you're not alone. I thought that, too. I *still* think that. After all, this isn't a fight about morals or emotions or politics; it's numerical and philosophical. It's about the *nature of truth.* And so I expected that one side would just lay down their metaphorical arms and agree with the other, based on the data.

Clearly, I was way too naïve.

Once I had talked to enough epidemiologists, I became appropriately jaded. One of them summed up perfectly how entrenched the parties are: "I've known Walt for thirty-five years. We've disagreed for thirty-five years."

But then I came across something that totally changed my perspective: it's called "adversarial collaboration," and it just means "people who violently disagree work together." Now, this is not like Democrats and Republicans working together, because that usually involves them setting aside their differences on a particular issue (for example, taxes) to work together on a less divisive issue (building roads). Instead, adversarial collaborations in science are when scientists work together *on the exact topic they disagree so violently on.* In other words, Willett and Ioannidis could work together on nutrition and health. You might reasonably ask: *Wouldn't that be like the Catholic Church collaborating with atheists?*

No, and here's why.

Ioannidis and Willett disagree on many things, but they agree that nutrition—and lifestyle more generally—is important, is worthy of study, and impacts our health. And that can be enough common ground for an adversarial collaboration.

Now, let me just say the obvious here: this is a tough sell. Science feuds can get pretty nasty pretty quick, and it takes a lot of pride swallowing on both sides to come together in good faith. But it is possible, and in my view it has some key benefits. First, you might actually learn something. Second, you might teach the other side something. And third, if you and your adversary are coauthors on the same paper, there's no way they'll criticize it when it's published.

It's entirely possible that a collaboration might only produce a paper that outlines what the disagreement is, what both sides' positions are, and what experiments could resolve it. Even that would be extremely valuable. Why? Well, many scientific arguments boil down to:

"You said X."

"No, I said Y."

"I distinctly remember reading X in your paper."

"Did you actually read the paper? I clearly said Y."

"Maybe if you spend less time pontificating and more time on your syntax, I could decipher your prose without an Enigma machine . . ."

This, of course, takes place over the course of years, in dueling letters to the editor. As I write this, Willett, Ioannidis, and a number of other scientists are having exactly this fight over the meaning of one sentence about twelve hazelnuts. So even the simple act of getting the parties in the same room to agree on what they disagree on would be progress.

When I talked to Ioannidis and Willett on the phone, I asked each of them whether they'd be willing to work with the other. And . . . both said yes! Sort of. One of them suggested it without my even asking; the other said "It's a possibility." Compared with people in government, that's basically a lovefest. I hope they do find a way to work together. It would be better for the rest of us.

If this were a Malcolm Gladwell book, right now you'd be reading a blistering rebuke of science as a whole. He'd use a story about a spaghetti sauce scientist to argue that you can't trust a damn thing any scientist says. And you might think he'd have a point: after all, I've just shown you a seemingly overwhelming number of careless errors, statistical chicanery, and everything in between hidden in the hallowed pages of scientific publishing. And, yes, there's no question that all of these things are warts, blights on the body of science. But you know how the saying goes: visible warts are better than invisible ones. In what other field do its own practitioners openly and publicly debate its flaws, no matter how fundamental? In other words, the only reason nutritional epidemiology is having a reckoning right now is that scientists themselves decided to give it one.

Not everyone agrees that nutritional epidemiology is in crisis. Its proponents have fought and will continue to fight hard against the onslaught of Ioannidis. Other scientists will judge both sides and choose. Once the dust of the war settles, there will be winners and losers. One camp will continue to be published in *Science* and *Nature* and claim the title of Settled Science, like the epidemiologists who said that smoking causes lung cancer. The other side will fade, although there will be those who never change their minds or disappear. Eventually, even the winners will be defeated by the next change in perspective or wave of new data, and they, too, will fade away. Sound familiar? It's a war like any other, except it's being waged in public, right now, all around you; and you have access to the whole messy battlefield. This is why I have confidence in science: not because it's perfect, but because you can find the imperfections and judge them for yourself.

Speaking of which, let's talk about *how* you get your science . . . via the news.

Writers of Internet headlines seem to belong to one of two groups: (a) people who think it's possible to *exactly maximize every aspect of your health with 100 percent certainty*, or (b) people trying to make money off group (a).

So now every time I see a headline like

EGGS ASSOCIATED WITH 27% INCREASED RISK OF HEART DISEASE

what pops into my head is

CLICK HERE FOR A SIMPLE WAY TO AVOID DEATH ALSO WE HAVE GREAT DEALS ON SMALL KITCHEN APPLIANCES

Reading news about food and health is like standing on the bow of the *Titanic* . . . without Kate Winslet. You look down and suddenly see a bit of ice floating in the water. Does that bit of ice extend down for hundreds of feet, cluing you in to a potentially deadly iceberg—or is it just an ice cube trying to sell you a toaster? Now imagine there are hundreds or even thousands of bits of ice ahead and you're surrounded by twenty-six people, each yelling at you to turn the ship and avoid *their* particular bit of ice, because *that one* is definitely an iceberg! Sometimes these twenty-six yelling people are random bloggers trying to sell you supplements; sometimes they're journalists exaggerating findings to get those all-too-precious clicks. Sometimes it's the scientists' institutions that exaggerate, puffing up press releases to get stories in major media outlets. And sometimes it's the scientists themselves who do it for tenure, or a higher profile, or simply because they're not skeptical enough of their own results. And, of course, sometimes there really is an iceberg out there. Smoking was a gigantic deadly iceberg.

Doctors and scientists are not immune to the onslaught. Richard Klausner, the former head of the National Cancer Institute, found himself at the bow of the ship back in 2001. "I'm pretty well plugged in to what's going on in research," he told Jerome Groopman of the *New Yorker*. "I hear on the news 'Major breakthrough in cancer!' And I think, Gee, I haven't heard anything major recently. Then I listen to the broadcast and realize that I've never heard of this breakthrough. And then I never hear of it again."

Most news about food and health just disappears into the night, bouncing harmlessly off the ship. And that brings us to my first piece of food advice in this whole book: Pay attention to safety alerts from the CDC or the FDA. Other than that, if you read something about food and health—especially individual foods, like kale and eggs—on the Internet, treat it like a kitten: have fun playing with it, but don't let it change your life. Don your Asshole Hat, try and poke a few holes, and then move on.

Why? Even if we give the news the benefit of the doubt and assume it'll report every scientific paper fully and faithfully, a single journal article is *not necessarily* proof of a fundamental truth.

Evidence takes years to accumulate—and even longer to become consensus. In short: a brick is not a bridge.

But, you might argue, isn't that all the more reason to pay attention to the headlines: so you can be alerted when Bridges of Truth are finally built? To which I'd say: No, because you don't read the news like scientists read science. Scientists are steeped in the literature of their fields. They've been reading it since grad school. They know all the key players. They (mostly) understand the pitfalls of the methods used. In other words, they have the right context. Regular people, like you and me, don't. For starters, we don't read the original journal articles; what we read has usually passed through at least a press officer and a journalist. But more important, we don't obsessively follow one topic. We haven't read about every association nutritional epidemiology has ever produced. All we do is dip our toes in the stream of news whenever we see a tweet that catches our eye or whenever our parents forward us an article. Remember the stream of headlines about coffee in the years before 2000? Imagine if you had read only three of these stories at random: you might think that coffee lowers your risk of hip fractures, increases your risk of lung cancer, and increases your risk of a heart attack. If, however, you'd been deeply steeped in the coffee literature of the past quarter century, you'd know that the results were bouncing all over the place and you'd probably have attached a little less weight to each individual one. You would also have seen a review of reviews published in 2017 that pooled the results from hundreds of studies over the years and found that many of the associations reported in those scary headlines just . . . disappeared.

Which brings me to the following spectrum:

NUTRITIONAL EPIDEMIOLOGY IS A USELESS, BARREN WASTELAND. ⟷ **NUTRITIONAL EPIDEMIOLOGY IS FINE; WHY ALL THE FUSS?!**

Where do you fall? Maybe slightly to the left of center? Maybe way to the right. If you think nutritional epidemiology is the bee's knees and Björk's fjord—that there are no problems with it whatsoever—fair play to you; I respect your opinion. But stick around for one more chapter, because there's something important we haven't covered yet:

One day, you will die.

So WHAT DO I DO?

This chapter is about how you should live your life . . . no pressure.

If you are a woman living in the United States and today happens to be your thirty-third birthday, *CONGRATULATIONS!* Happy birthday! The probability that you will die before your next one is about 0.0884 percent, or about 1 in 1,131. If you're a man, the equivalent probability is 0.175 percent, or about 1 in 571. How the hell could I possibly know that? Turns out, it's an easy calculation once you have the right data.

In 2017, 2,813,503 Americans died. In 2016, that number was 2,744,248. In 2015, it was 2,712,630. Almost every single death in America is categorized and counted by the Centers for Disease Control and Prevention, yielding an incredible amount of data that government scientists spend years analyzing. Throw in a pinch of statistics and a splash of calculus, and you can roughly estimate the risk of death for the average American man and woman. Every year the CDC publishes these estimated risks in what's called a "life table," and the beating heart of a life table is two columns of numbers that look like this:[*]

....

[*] If you ever decide to peruse life tables on your own, keep in mind that the risk of death is usually expressed as a probability ranging from 0 to 1; to convert to a percentage, you just multiply by 100. So for example, the

YOUR AGE ↓   YOUR (APPROXIMATE) RISK OF DYING WHILE THIS AGE (AVERAGE OF MEN & WOMEN)

Age	Risk	Age	Risk	Age	Risk	Age	Risk
0–1	0.5894 %	25–26	0.1004 %	50–51	0.4098 %	75–76	2.9614 %
1–2	0.0403 %	26–27	0.1028 %	51–52	0.4481 %	76–77	3.2507 %
2–3	0.0252 %	27–28	0.1056 %	52–53	0.4885 %	77–78	3.5786 %
3–4	0.0193 %	28–29	0.1094 %	53–54	0.5319 %	78–79	3.9616 %
4–5	0.0145 %	29–30	0.1138 %	54–55	0.5781 %	79–80	4.4017 %
5–6	0.0143 %	30–31	0.1185 %	55–56	0.6271 %	80–81	4.8899 %
6–7	0.0128 %	31–32	0.1232 %	56–57	0.6745 %	81–82	5.4283 %
7–8	0.0116 %	32–33	0.1277 %	57–58	0.7291 %	82–83	6.0367 %
8–9	0.0104 %	33–34	0.1318 %	58–59	0.7824 %	83–84	6.6954 %
9–10	0.0095 %	34–35	0.1359 %	59–60	0.8383 %	84–85	7.4533 %
10–11	0.0091 %	35–36	0.1408 %	60–61	0.8991 %	85–86	8.2695 %
11–12	0.0098 %	36–37	0.1468 %	61–62	0.9652 %	86–87	9.2575 %
12–13	0.0125 %	37–38	0.1535 %	62–63	1.0353 %	87–88	10.3427 %
13–14	0.0174 %	38–39	0.1608 %	63–64	1.1081 %	88–89	11.5296 %
14–15	0.0241 %	39–40	0.1690 %	64–65	1.1838 %	89–90	12.8216 %
15–16	0.0314 %	40–41	0.1790 %	65–66	1.2634 %	90–91	14.2211 %
16–17	0.0388 %	41–42	0.1909 %	66–67	1.3510 %	91–92	15.7287 %
17–18	0.0473 %	42–43	0.2043 %	67–68	1.4504 %	92–93	17.3433 %
18–19	0.0566 %	43–44	0.2191 %	68–69	1.5664 %	93–94	19.0616 %
19–20	0.0660 %	44–45	0.2360 %	69–70	1.7059 %	94–95	20.8781 %
20–21	0.0757 %	45–46	0.2541 %	70–71	1.8766 %	95–96	22.7849 %
21–22	0.0846 %	46–47	0.2752 %	71–72	2.0689 %	96–97	24.7715 %
22–23	0.0914 %	47–48	0.3018 %	72–73	2.2709 %	97–98	26.8255 %
23–24	0.0958 %	48–49	0.3346 %	73–74	2.4795 %	98–99	28.9322 %
24–25	0.0984 %	49–50	0.3717 %	74–75	2.7078 %	99–100	31.0753 %
						100 AND OVER	100.0000 %

The beating heart of a life table is risk of death. Setting aside the irony, we can actually learn a lot from this unassuming list of numbers. For one thing, looking at the table above should immediately confirm whether you're an optimist or a pessimist: Do you, thirty-three-year-old birthday woman, see the small but not-quite-zero 0.0884 percent risk you'll die during your thirty-third year, or the 99.9116 percent (almost) guarantee you will survive to your thirty-fourth?

Another interesting thing to note is that, starting around your twenties or thirties, the risk of death starts increasing by roughly 8 percent per year, meaning that if you take last year's risk and multiply it by 1.08, you will get this year's risk. This seems pretty

....

probability that you will die while thirty years old is 0.001185, which is the same as 0.1185 percent. In the table above, I've already converted everything to percentages.

modest. But let's briefly travel back to 1986, when interest rates for certificates of deposit were also in the 8 percent range. (I know this is a bit weird, but stick with me here.) If a bank back in 1986 had offered an 8 percent certificate of deposit with a fifty-year term, and if you had deposited $10,000 into that account, how much money would you have at the end of the term? You might think, *Eight percent of $10,000 times fifty years,* which would give you about $40,000. You'd be wrong. At the end of fifty years, you'd actually have more than half a million dollars. This is why your parents—and that guy who screams at you on CNBC—want you to save money: because of the power of compound interest, also known as the power of *time.* What's even more amazing than how *much* you earn over the fifty years is *when* you earn it. The first year after you make the initial deposit, you earn about $800; the fiftieth year, you earn $41,000. In other words, you earn the most near the end.

Replace the word "earn" with "die" in the previous sentence and you get a statement that is just as true: you die the most near the end. What works *for* you with money—time—works *against* you with death. In fact, the underlying math is the same in both cases; it's what's known as an "exponential" increase, meaning that your risk of death increases more quickly the older you get.[*] Alyson van Raalte, a demographer who helped me decipher life tables, offered this gloomy tidbit in a cheery tone: "Most people don't realize just how quickly our death rates increase with age."

I *definitely* didn't realize. When I started poking and prodding the numbers in the CDC's life table, I terrified the crap out of myself. For example, an eighty-five-year-old's risk of death while eighty-

....

[*] This exponential function holds until about age 105. After that, we don't really know what happens . . .

five is *912 times* that of a ten-year-old's risk of death while ten. That's 91,200 percent as much! Even when you compare an eighty-five-year-old to a fifty-year-old, the numbers are wild: an eighty-five-year-old's risk of death while eighty-five is 2,020 percent as high as a fifty-year-old's while fifty!

So far the pessimists out there are like "YEAH! SEE, EVERY-THING *IS* SHIT!" But wait, there's more. The most striking thing about a life table is that the risk of dying at any given age is low. It's also lower than you would expect it to be. For example, a forty-year-old male's risk of death during his fortieth year is only 0.224 percent. A fifty-year-old American woman's risk of dying while fifty is only 0.320 percent. These risks seem even lower when you consider everything that's trying to kill you. (By the way, despite Australia's reputation as Land Where All Objects, Animate and Inanimate, Are Conspiring to Bring About Your Imminent Death, a forty-year-old Aussie's risk of dying while forty is actually 0.142 percent, lower than a forty-year-old American's.)

Here's a question: At what age do you think an American first faces a 10 percent risk of death in the next year? In other words, at what stage in your life do you think the risk of dying within a year first becomes one in ten?

Sixty?

Seventy?

Eighty?

Nope. Eighty-seven.

Remember that eighty-five-year-old whose risk of death is 91,200 percent as high as a ten-year-old's? Well, that old fart also has eleven to one odds of surviving to their eighty-sixth birthday; their risk of death that year is "only" 8.27 percent. The CDC tables provide detailed probabilities only up to age 100; the probability of death within that year is only 34.5 percent. In other words, if you get to 100, there's roughly a two-thirds chance you'll survive to 101. Even to me, a wholehearted pessimist, that seems pretty damn optimistic!

But wait—the optimists haven't won. Things start to look gloomy again if, instead of looking at it from the perspective of one year at a time, you look at ten-year groups. For example, let's go back to our forty-year-old American man. The probability that he will die during his fortieth year is only 0.224 percent. But the probability that he will die sometime within the ten following years is 3.2 percent. At age fifty, that jumps to 7.4 percent. At age sixty, it's 15 percent. At age seventy, it's 31 percent. And at age seventy-five, it's 45 percent. So at age seventy-five, flip a coin to see if you'll still be alive by age eighty-five.

Some would see that as depressing; others would see it as pretty amazing. But I'm not here to depress or amaze. I'm here to divide . . . numbers.

One thing that stands out when you look at recent American life tables is that men and women share the age at which they're *least* likely to die: ten. Ten-year-olds have a 0.0091 percent risk of dying before their eleventh birthdays, and a 99.9909 percent risk of living to see what the Kardashians will do next. Because a ten-year-old's risk of death (while ten) is so incredibly low, almost any number you divide by this will seem gargantuan. For example, if you divide the risk of death while age twenty by the risk of death while age ten, you get a factor of 8, meaning that a twenty-year-old is 8 times as likely—or 800 percent as likely—to die as a ten-year-old. Here's where the line between optimism and pessimism gets blurry: it's true that a twenty-year-old's risk of death is incredibly low; it's *also* true that it's *much higher* than a ten-year-old's.

Exponential functions, like compound interest or risk of death, are notoriously tricky. It's hard to keep a sense of scale of the entire function in your brain at once; if you understand one side of the function (risk of death in your early thirties, or growth of money in the first few years), it's really hard to wrap your head around how much higher the *other* end of the function is (vast fucking wealth . . . or death). We *do* have a decent intuitive understanding: when a ten-year-old dies, it's shocking, as it (mathematically and emotionally) should be; when a ninety-nine-year-old dies, it's sad but not unexpected. But that intuitive understanding does not translate into a mathematical understanding. Percentagewise, the difference between the minimum risk of death (at age ten) and the maximum (at age ninety-nine) is much bigger than we expect: over 340,000 percent. Three hundred and forty thousand percent![*]

That raises the question: How much can your intake of ultraprocessed foods kill you?

To answer that, let's look back at the study that linked ultraprocessed foods with an increased risk of death. The authors found that for every 10 percent increase in the proportion by weight of ultra-processed food in the participants' diet, their annual risk of death went up by 14 percent. This result comes from the same folks who used the French NutriNet-Santé participants to associate ultra-processed food with cancer. You can—and we did, in chapters seven and eight—list a few reasons to be skeptical of this number. But let's stop with the "Well, actually . . ." routine for a minute. Let's pretend that eating ultra-processed food truly did increase the participants' annual risk of death by 14 percent; in other words, let's pretend that we know this association to be legit and causal.

....

[*] Conclusion: Spend a little more time with the old fogies in your life. Their risk of death is a lot higher than yours . . . and it's increasing a lot faster, too.

Sounds *terrible*, right? If I were a headline writer, I might inject a little flair (and inaccuracy) by writing something like:

GOD HATES PEOPLE WHO EAT CHEETOS, SHORTENS LIFE BY 14% STUDY FINDS

If a 14 percent increase in the risk of death translated to a 14 percent reduction of your total life span, that would be . . . *awful!*[*] Fourteen percent of the average American life span is about eleven years. That's a lot of life lost. But it turns out that "increasing risk of death" and "shortening life expectancy" are *not* mathematically the same. To see why, let's do a quick back-of-the-envelope calculation. My risk of dying in the next year is about 0.18 percent. Now let's bump it up by 14 percent. My new risk of death for the year is about $0.18 \times 1.14 = 0.21$ percent. If we then flip the script and look at risk of *living* and plot these numbers on a graph, we get:

....

* For complicated mathy reasons, a 14 percent increased risk of death would technically correspond to a $1 - 1/1.14 = 12$ percent reduction in life span, but there's no need to nitpick.

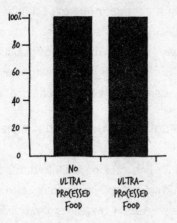

PROBABILITY I WILL LIVE TO NEXT YEAR

You can't see the difference. (I had to zoom in in Excel to confirm it's there. It is.) Even if I very pessimistically assume I'm going to eat 10 percent more ultra-processed food in my diet for the next ten years and compare my risk of living over the ten-year period between fifty and sixty, the result is pretty much the same:

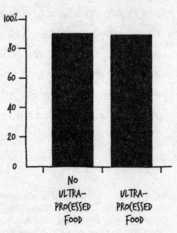

PROBABILITY I WILL LIVE FROM AGE 50 TO 60

If the authors of the study had wanted to convert the 14 percent increase in risk to a change in life expectancy, they could have used fancy math—stuff like "accelerated lifetime models." They didn't. And because the raw data is not publicly available, you and I can't either. But it turns out you can approximate the change in life expectancy using a simple mathematical equation:[*]

CHANGE IN LIFE EXPECTANCY ≈ −10 × ln(RELATIVE RISK)

The "ln" stands for natural log: you'll find this function on most calculators. So, in this example, if we really and truly believe that eating ultra-processed food increases your risk of death by 14 percent, the approximate change in life expectancy would be:

−10 × ln(100% + 14%) = −10 × ln(114%) = −10 × ln(1.14) = ABOUT 1.3 YEARS LOST

Whaa? How the hell can such a large-sounding change in the risk of death (14 percent off!) produce such a small effect (about 2 percent of an average American life span)? The culprit is our old friend, the exponential death function: specifically between the ages of ten and seventy, when the risk of death within any given year is low.

So from my perspective, when it comes to risk of death, the 14 percent increase associated with ultra-processed food—assuming it's real—is small potatoes. It *seems* like a lot, because we're so used to thinking of percentages as maxing out at 100 percent; but in the context of the relative risks of death we all face on a regular basis, 14 percent is low. It is, in fact, almost exactly the risk incurred by the simple act of . . . turning twenty. Remember

....

[*] And an important assumption: that the annual increased risk stays constant throughout your entire life. This isn't really true, but we'll just roll with it. And while you're down here: if you happen to be a professional statistician, you know that this isn't quite as simple as I'm making it cut to be. But it's good enough for the back of a cosmo-soaked napkin.

smoking? A study of about 35,000 British doctors found that heavy smokers were 234 percent as likely to die each year (from any cause) as nonsmokers. This is a much more impressive number, and it translates into a much more impressive reduction in life expectancy: about ten years on average.

But there are other very large risks besides smoking.

Being a man is one. In fact, in developed countries, there is no age at which a man's risk of death is lower than a woman's. At its peak, the risk of death ratio hits 2.85, meaning men are 285 percent as likely to die as women. (Unsurprisingly, this happens at age twenty-two. Demographers call this the "Accident Hump," which I like to think of as the "Young men are more likely to do dumb shit, but it's not their fault; it's the testosterone or maybe a social environment that encourages acts of daredevilism" Hump.) Your income is another risk factor. In the United States, people aged forty to seventy-six who are in the top 1 percent by income (as measured by the Social Security Administration) live ten to fifteen *years* longer than people in the bottom 1 percent. Where you live is yet another. The poorest Americans in New York City live, on average, four years longer than the poorest Americans in Gary, Indiana. As you might expect, race matters, too. For example, black babies under the age of one are more than *twice* as likely—231 percent more likely—to die as white babies.

You can't change your age. You can't change your race. And it's hard to change your income or where you live. But it's pretty easy to eat less ultra-processed foods. Or more "superfoods." Or switch to the Mediterranean diet. And changing what you eat is a lot easier than changing your income. Therein, I suspect, lies the lure of dietary changes: they feel easy, and with numbers like 14 percent, they make you feel like you're substantially affecting your risk of death . . . even though you're not.

I suppose I'm making a somewhat mathematical and heartless argument for something the Dude has been saying all this time: Relax, dude. Or, as a modern-day Twitter user would put it: lol nothing matters.

As I was writing these last chapters, I came across a recent paper from Walter Willett's group that looked at associations between five health behaviors (also known as lifestyle choices) and death . . . kind of like five classic nutritional epidemiology studies rolled into one. The great thing about this paper is that instead of just giving you scary-sounding phrases like "increases risk of death by 27 percent," Willett's group actually put the risks into a format we can all clearly understand: number of years of life gained or lost. Also, it's one of the few papers that analyzes five different health behaviors using the same math.

So I was extremely curious to see what nutritional epidemiology says you should do—if you trust its results implicitly.

What exactly did Willett's group do? Essentially, they calculated a bunch of risks of death in the traditional way (as relative risks) and then they calculated how those risks would affect the remaining life expectancies of fifty-year-olds. So, for example, they calculated that the risk of death of a heavy smoker (twenty-five or more cigarettes per day) was about 287 percent as high as the risk of death of a never-smoker. Then they took that number and sprinkled it onto life tables for Americans aged fifty years and over and, like salt on lettuce, watched it suck the life right out of the life table. Basically, they calculated how soon you could expect to die if you were a fifty-year-old current 2.5 pack-a-day smoker vs. a fifty-year-old never-smoker. And the answer was: about twelve years earlier if you're male, and nine if you're female.

For each lifestyle choice, they mathematically split up the participants into groups, corrected for some confounding variables, and compared the life expectancies of each group to the others.

What did they find for physical activity?

People who did 3.5 hours or more per week of "moderate or vigorous physical activity" lived roughly eight years longer than those who did no exercise at all, but even as little as 0.1 to 0.9 hours per week gave you an expected survival boost of five years.

Obesity?

People who were class 2 or 3 obese (having a BMI of over 35) had a life span between four to six years shorter than those with a BMI of 23 to 25. People with a BMI of 25 to 30 had a life span of only a year shorter than those with a BMI of 23 to 25.

Alcohol?

People who didn't drink at all and people who drank 30 grams of alcohol per day had roughly the same life span—about two years shorter—than people who drank between 5 and 15 grams per day.

And lastly, diet.

People who ate the "healthiest" diet had a four- to five-year-longer life expectancy than people who ate the least healthy diet. That probably raises a bunch of questions; don't worry, we'll get to them all, but first I want to point out a couple key things.

First, remember that all these numbers are based on observational data (no randomized controlled trials here) and all the life expectancy calculations assume that the change in life expectancy is *caused* by the lifestyle choice, not just associated with it. In other words, we're assuming that the associations are legit and causal.

Second, according to this analysis, the gains in life expectancy are . . . huge! If you compare people in the absolute worst groups in every lifestyle choice to those in the absolute best, the life expectancy difference is about twenty years! And remember, that's at age fifty. So it's the difference between living to ninety-four or seventy-four.

I had many feelings when I first saw this. My first reaction was: *Well, that sounds wronger than pickles in a piñata!*

But then I got to thinkin' . . .

The twenty-year difference came from comparing people with *every single health trait maximized to perfection* to those with *every health trait minimized to disaster*. In other words, you're comparing healthy-weighted people who never smoke, drink moderately, exercise their butts off, and eat the "healthiest diet" to morbidly obese chain-smoking alcoholics who eat the "worst" possible diet and do less than six minutes of physical activity per week.

My second reaction was: *Well, when you put it like that, a twenty-year difference in life span doesn't . . . actually . . . seem . . . that . . . crazy.*

Spelling out who exactly is being compared brings up another important point: the people in the absolute worst and absolute best health groups are rare. Using data from the CDC, Willett's group estimated that only 0.14 percent of Americans were in the worst health group and 0.29 percent were in the best. Most people were pretty tightly huddled around the middle. You could look at this important fact in one of two ways. Optimists might think, *Wow, 99.86 percent of Americans can improve their lifestyles! Much opportunity! Very disrupt!* Pessimists might look at these same numbers and think, *Oh, only 0.29 percent of Americans have maximized five separate and difficult dimensions of their lifestyle? Congratulations on winning this year's No Shit Obviously Duh Award!*

When you start looking at what would be required to tack on a significant number of years to your life span, things look . . . yeah, still pessimistic. For example, if you wanted to add about three years to your life expectancy at age fifty, you could:

- **lose 5 BMI units (that's a lot)**
- **cut back from 20 cigarettes a day to 10**
- **do 4 hours of physical activity per week instead of 2**

To be clear, you'd have to do *all* these things, not just one of them.

That's . . . hard!

If you change your perspective and look at *all Americans* instead of just one person, things look a bit more optimistic. For example, if doubling your physical activity from two to four hours per week gives you-the-individual another year of life, eh, who cares? But if 10 percent of the American population doubles their physical activity, that's a lot of extra kids' birthdays that get seen. Plus, it probably has the added bonus of reducing the strain on our already very strained health care system. All of this is assuming, of course, the association is legit and causal.

So, given all that, what would I advise you to do?

———————

I hereby present:

Four Bits of Advice, Assuming You Are Generally in Good Health

However, please note: I'm not a doctor. If your doctor tells you to do something that contradicts anything in this book, follow your doctor's advice and ignore mine. Your doctor knows you; I don't.

Okay, that said, here we go . . .

Bit of Advice #1

Don't worry so much. Ignore most news about food and health— unless it's a safety recall or contamination notice or something like that. Health news is not designed to give you contextualized, nuanced information about your health; it's designed to sell advertising and cookbooks. It's the *latest* word, not the *last* word.

If you want honest-to-goodness information about specific

foods, diets, or drugs, the best and easiest place to get that information is the Cochrane Database of Systematic Reviews. They review studies on many—but not all—foods, interventions, and health outcomes, and they try to detect and call out the chicanery that makes its way into the scientific literature. Cochrane isn't perfect, but they compress the most amount of information and the most rigorous review of the strength of the evidence into the shortest and easiest-to-read summary. Plus, they constantly update their reviews as new science is published.

Bit of Advice #2

Don't smoke. If you do smoke, quit. If you're a smoker, quitting is the single most important thing you can do to live longer. If you don't smoke, not starting is the single most important thing you can do to live longer.

What about vaping? If you're already a smoker, there's evidence that vaping can help you quit, and it may be just as effective as (or more effective than) nicotine replacement therapy. That said, if you don't already smoke, there's literally no good reason to start vaping—partly because we already know it contains some of the same carcinogens as cigarette smoke, but also because it could be an on-ramp to smoking.

Bit of Advice #3

Be physically active. It's not as clear as it is with smoking whether physical activity *causes* you to have a longer life or whether it's just *associated* with a longer life. But the distinction doesn't matter. Physical activity feels good and has basically no risks,[*] so you might as well do it.

....

[*] If your physical activity of choice is crocodile wrangling, you will have an appreciably higher risk of death than if it's "brisk walking," which is what

Bit of Advice #4

This is the one about food. Why is this last on my list? Remember that, in Willett's analysis, people with the healthiest diet had a four- to five-year-longer life expectancy (at age fifty) than people with the least healthy diet. But what did the healthiest diet look like? It had lots of fruits, veggies, nuts, whole grains, polyunsaturated fatty acids, and long-chain omega-3 fatty acids, and very little or no processed meat, red meat, sugary drinks, trans fat, and salt.

Let's put these eleven items on a to-do list. Notice anything about that list? I notice two things.

First, it's long. (By comparison, "Don't smoke" and "Be physically active" are one item each.) Why does that matter? Because if eleven items *together* add up to roughly 4.5 years of life expectancy gain, each individual item would contribute only about five months (assuming the items contribute equally).

Second, the list is actually much longer than eleven items. It's actually four individual items (salt, trans fat, polyunsaturated fatty acids, and omega-3 fatty acids) and seven *entire categories of foods* (fruit, veggies, nuts, whole grains, processed meat, red meat, and sugary drinks). Even if you charitably count "sugary drinks" as one item, you've still got five actual food items and six categories, each of which has tens or hundreds of foods in it. Plus, the five "food items" are actually fairly specific molecules (salt, sugar, fatty acids, and trans fats) found in a wide variety of foods. My point here is this: while "a healthy diet" may sound like One Big Thing, it's actually a fairly complicated mishmash of hundreds of little things. And that means that the contribution of any one food item (like blueberries or dark roast coffee) to life expectancy is probably very, very small.

....

typically counts as physical activity in scientific studies. Also, if you're really out of shape, physical activity can be riskier.

None of this means you shouldn't try to eat a healthy diet. My point is that you shouldn't kill yourself with worry about which species of fish is highest in polyunsaturated fatty acids, or whether a ripe avocado has fewer omega-3s than an unripe one, or whether light or dark roast coffee is higher in antioxidants, or any other poppycock I've seen in health magazines. And for this same reason, it probably doesn't matter too much exactly which diet you pick: almost any non-snake-oil-salesman-and-real-doctor-recommended diet is going to check most of the right boxes, and even if it's missing one or two or has an extra one, it will make almost no difference to your life expectancy.

Of course, diets are not just about living longer. Sometimes—maybe even most of the time—they're about "being your best you," or as non-millennials would say it, feeling better. You've almost certainly experienced the feeling of going on a diet and then feeling healthier or better in some way. Problem is, there's no way to tell whether you feel better because you're going on *this* specific diet or because you're going on *any* diet. The simple act of going on a diet (*any* diet!) could make you feel better just by itself. Also, if you're going on a diet, you're probably also exercising more, spending less time hungover, sleeping more, etc. All these things could make you feel better, independently of the diet.

What about cutting out ultra-processed foods from your diet completely? You might ask: Is there enough high-quality evidence to justify doing that? Do we have a solid Bridge of Truth between ultra-processed foods and death, like we do for smoking? Absolutely not. But then again, the surgeon general didn't wait around for an unassailable bridge to be constructed before telling people to stop smoking. If you've read the evidence in this book and your reaction is *Hey, better safe than sorry,* you would have a lot of company. After all, there is no known risk associated with *not* eating Cheetos—or any other ultra-processed foods—so why not just cut them out of your diet entirely?

Many dietary recommendations say exactly this: Avoid processed food. And I have to say, I don't disagree with that fundamental point. But . . . can we please stop calling it a poison? It's disrespectful to the *REAL HONEST TO GOD* poisons out there, which work really hard at giving you diarrhea or stopping your heart. When we start calling things like sugar (or even ultra-processed food) poison, it cheapens the term. No one would say candy is good for you, but it's not cyanide.

You might think I'm being pedantic, and I am. But consider this: if you literally believe that ultra-processed food is a horrible poison, you might logically conclude that if you cut out ultra-processed food from your diet, you can keep chain-smoking and the risks will cancel each other out. That is wildly delusional. Also, if everyone keeps saying things are poison, we'll all get so jaded that when a *real* poison comes along, we'll ignore it, like a boy-who-cried-food type situation. So, if you decide to cut out all ultra-processed food, that's totally fine: it might make you feel better, whether from the placebo effect or, just as likely, because you'll have to replace all the ultra-processed food with fruits, veggies, and other stuff most diets would tell you to eat anyway.

But the one thing cutting out ultra-processed food will *definitely not do* is give you eternally long life.

EPILOGUE

This epilogue is about Velveeta, personal responsibility, and kielbasa.

We want food to be like the Harry Potter stories, where we all know that Dumbledore is unblemishably good and Voldemort is irredeemably evil, but instead food is more like a French art house film called *La fin des haricots*: everyone's flawed and you're not even sure what happened.

Overall, though, I find myself agreeing with John Ioannidis more than Walter Willett. Yes, prospective cohort studies have some strengths. They give us long-term data from people just living their regular lives, not being required to eat a particular food. They can generate potentially interesting associations to test in randomized controlled trials. And in certain cases, like with smoking, you can be pretty sure an association is causal. But from all the research I did, I'm just not convinced that these types of studies are reliable and accurate enough to detect things like a 14 percent change in risk. I also think it's way too easy for everyone—scientists, journalists, and regular folks—to see an association between two things and assume that one causes the other. Even if I give traditional nutritional epidemiology the benefit of the doubt and assume that the 14 percent increase in mortality risk is both legit and causal, that's pretty low: only about a year of life lost.

Remember at the beginning of the book when I said this

journey had wholly changed my perspective on food and all the stuff we call "consumer products"? It certainly has, but it's also done more than that. It's made me see science in a whole new—and better—light. That seems weird, I know. After all, I *did* just spend the last ninety pages telling you about all the surprising potholes on the road to a legit and causal association. But the most important thing I learned is also probably the most obvious: uncovering the truth about all the stuff we eat, drink, inhale, and smear on ourselves is much harder than it seems. The world isn't usually like intro organic chemistry, with clean, simple reactions leading to clean, simple products. It's more like advanced organic chemistry, in which All Hell Breaks Loose. Even if you manage to uncover the truth, sometimes that truth is complicated. If ultra-processed food really does turn out to increase the risk of death by 14 percent, that'll raise almost as many questions as it answers: Are all ultra-processed foods equally bad? What exactly makes them bad? Can they be changed to make them less bad or even good?

Science proceeds slowly and erratically. If you're on the outside looking in, trying to figure out what's true can be insanely frustrating. But once a super-solid Bridge of Truth is constructed, it's a beautiful thing, just like the process that created it: science.

Throughout this book, I've been treating science as if it operated separately from all the normal human concerns, like the influence of corporate money and power. Of course science doesn't operate in a vacuum. If you've paid any attention to the food industry's response to the many food movements of the past fifteen years, you might notice that their arguments bear some striking similarities to mine.

For example, "Eat reasonably and don't worry so much" is another way of saying, "We companies shouldn't be regulated on what we can sell; the regulation should be in your own heart, at the individual level, when choosing what you buy and eat." In a similar vein, you could interpret "Exercise more" as a veiled way of saying, "To compensate for the sin of eating these crappy ultra-processed foods, *you the consumer* have to do a bunch of work, *not us.*"

American society prizes personal responsibility, so we're usually open to these types of arguments. But it's also wildly disingenuous for an industry to make an addictive product and then turn around and tell us that it's *our* responsibility to resist the addiction. So your beliefs here might just come down to whether you believe ultra-processed foods are addictive. To me, they definitely are. Most of us probably have one guilty-pleasure food-in-a-box that we can't stop ourselves from eating. (Mine is chocolate Necco wafers. I know, it's weird.)

But even if we prove ultra-processed food is both addictive and bad for you, it'll probably stay on supermarket shelves. After all, we know cigarettes are addictive and multiply the risk of lung cancer by 11 or more, and they're still being sold decades later. Even if ultra-processed food turns out to be *the only cause* of the obesity epidemic, could you imagine a law that tried to ban all ultra-processed foods? Fox News would air a segment called *V for Velveeta* and it would die in the Senate faster than a gastrotrich.*

That said, there are those who say that taxing certain types of ultra-processed foods, like soda, is clearly the way forward. I think there's a case to be made for taxing soda, because it has zero nutritional value. But taxing *all* ultra-processed foods seems far less likely, especially because they're cheap. If you have trouble making ends meet, they could be cheap enough to prevent you or your family from going hungry. Remember Kevin Hall's randomized controlled trial? Based on how much it cost to prepare meals for that study, Hall estimated that the cost of 2,000 calories was about $15 for ultra-processed food vs. $22 for

....

* Gastrotrichs are tiny marine animals that live only a few days.

minimally processed food. That's a difference of about $2,500 per person per year. If you've got a family of four, that's ten grand a year. So for many Americans cutting out ultra-processed food is not a choice; it's a luxury.

I don't see a simple solution here, especially because at some point it'll probably have to make its way through our political system . . . one of the few irreplaceable ingredients of our lives.

But back to you.

If you still find yourself agreeing with the Concerned Consumer at the beginning of this book; if you have deep and existential worry about food and toxicity and chemicals—if you'd rather eat the nutritional-epidemiology-certified "healthiest" diet just to be safe—I probably haven't changed your mind about the underlying science. But even if you fully believe that potholes on the road to legit and causal associations are no big deal, changing your diet probably won't change your life expectancy by more than a few years.

Is that worth it?

That's up to you.

For me, right now, getting 1.3 extra years by eating 10 percent less ultra-processed food doesn't feel worth it. But maybe that's because I'm young. From the vantage point of 33 years old, the difference between 77 and 78.3 seems small. But if I were at death's door, things might feel different. One demographer I talked to put it like this: imagine being told you're going to die tomorrow vs. being told you're going to die in fifteen months. That same 1.3 years feels a lot longer.

But I'm not even sure that "1.3 years," or "14 percent" is legit and causal.

I interviewed a lot of scientists with wildly different views on whether nutritional epidemiology is horse caca. Then, as I was writing this epilogue, I stumbled upon the chance to actually talk to one of the *participants* of a long-term prospective cohort study, who generously donated data about their life in the hopes of making others' lives better. This person had been a regular participant in a fairly prominent study for years, and I wondered what they would say about the problems of memory, or confounding variables, or p-hacking, or statistical skulduggery. As it turned out, they said nothing about any of that. I asked them whether they had changed their diet because of the results of the study. They answered:

> I don't see myself as the sum total of what I ingest. My mood and my company and a bunch of other things have a lot to do with my health. So I never stopped eating butter. I never stopped drinking wine. I had sugar every day. I figure if what I eat makes me happy, then it's just fine. But I do make shifts. I do respond gradually. I don't eat nearly as much kielbasa as I used to.

That seems like the right approach to me.

I also think there are more important things to worry about than how food changes your life expectancy. Things like climate change and the sudden dip in popularity of vaccinating ones' children, I'd argue, will have a much bigger impact on people's lives.

But that's another book.

APPENDIX

How Does Hand Sanitizer Work?

Let's talk about the consumer product that promises to de-germ your hands no matter how far you are from a sink: hand sanitizer. What exactly does it do? And does it really work? Knowing *whether* something works and *how* it works are, as we've seen throughout this book, two different things. Let's look at *whether* first. To do that, we have to travel more than a quarter-millennium back in time, to eighteenth-century Vienna, Austria. *Wilkommen!*

In 1784, Austria's Vienna General Hospital opened what we would now call a maternity ward. Over the next four decades, 71,395 babies were delivered. In 897 of those deliveries, the mother—unfortunately—passed away. If you divide the number of maternal deaths by the number of births, you get what's called the "maternal mortality rate," an indication of how dangerous childbirth is for the mother. From 1784 to 1822 at the Vienna General Hospital, the maternal mortality rate was roughly 12.6 deaths per 1,000 births.*

....

* It's a lot safer to give birth today than it was in the late eighteenth century. In the UK, maternal mortality these days is less than ten deaths per 100,000 births—that's about a hundred *times* lower (10,000% lower risk) than two hundred years ago.

Here's where things start to get weird.

In 1823, the hospital found itself under new management. Under the previous director, autopsies on dead mothers were not performed routinely; under the new director, they were. Over the next ten years, maternal mortality *quadrupled*, to 53 deaths per 1,000 births.

Well, shit.

But hang on a second. If you've learned anything from this book, you know that what I'm describing here is an *association*. The higher maternal mortality rate was *associated* with autopsies, but no one had actually set up a randomized controlled trial.

Here's where things get even weirder.

By 1833, the maternity ward was so overcrowded that the Folks In Charge decided to build an extension. Where there had been *one* clinic, now there were *two*. And then, in 1839, the Powers That Be decreed that one clinic be used *only* to train medical students, and the other *only* to train student midwives. Expectant mothers were randomly assigned to clinics, ensuring that the women in each clinic were roughly the same in terms of wealth, health, and other variables.

By sheer accident, the Vienna General Hospital had set up a randomized controlled trial. And, like any good randomized controlled trial, only one thing differed between the two clinics. In the midwife clinic, there were neither autopsies nor routine vaginal examinations of pregnant women. In the medical student clinic, the doctors-in-training started their day with autopsies on mothers who had died the previous day, and then—*without washing their hands*—performed vaginal exams on women about to give birth.

For fuck's sake, right?*

....

* Sources conflict on whether the medical students washed their hands after autopsies. Some claim they didn't; others suggest they did, but that even after washing, the medical students' hands still smelled like dead body.

This accidental trial lasted eight years, from 1839–1847. Results? In the midwife clinic, the maternal death rate was 33.8 per 1,000 births. In the medical student clinic, it was almost *triple* that—90.2 deaths per 1,000 births. Needless to say, that's a *noticeable* difference, to the point where some women chose to give birth in the street rather than be admitted to the medical student clinic.

In 1846, a twenty-eight-year-old doctor named Ignaz Semmelweis joined the medical student clinic as an assistant to one of the professors. He immediately noticed the shocking difference in mortality rates and tried to figure out what the hell was going on. Unfortunately, the breakthrough was a tragic one. In 1847, his doctor buddy Jakob got nicked with a scalpel while performing an autopsy. He died soon after, and *his* autopsy was strikingly similar to those of the mothers dying in the clinic. Semmelweis immediately proposed that the medical students were (unintentionally) contaminating their hands with what he called "cadaverous particles" and transferring them directly into women's vaginas.

But he went further than just proposing an explanation. He proposed a solution. In May of 1847, he started requiring that medical students disinfect their hands after autopsies using a solution of "chlorinated lime." (Today, this is called calcium hypochlorite, and you probably have a close chemical cousin tucked away somewhere in your house: bleach.) What happened next? You guessed it: maternal mortality fell to 12.7 deaths per 1,000 births, right in line with the rate in the midwife clinic (13.3 deaths per 1,000 births).

Semmelweis was not the first person to champion handwashing, but he was the first to prove, using modern scientific methods, that using a hand sanitizer could save lives. (Incidentally, he figured this out decades before the medical community accepted that germs, and not "foul air," could cause diseases. These days we know that there are many different types of unimaginably small life forms. Bacteria, viruses, and fungi are three common ones; most are harmless, but there are a few bad actors within each category that can infect us and wreak all kinds of havoc. We call the bad actors "germs.")

Technically, you could still use bleach as a hand sanitizer, but I wouldn't recommend it. Prolonged contact (especially with concentrated bleach) can be painful, give you blisters, or even kill skin tissue. And, even though a certain powerful American politician has hinted at drinking it . . . I wouldn't. Bleach isn't instantly deadly in tiny quantities (like cyanide is), but people *have* died from drinking it. Incidentally, I also wouldn't inject bleach. (One woman wound up in the intensive care unit after injecting 100 milliliters of bleach into her jugular vein. Frankly, it's amaziing she lived.)

So even though bleach may have been the canonical hand sanitizer, it's not the active ingredient in what we use today. That title belongs to the very same molecule that gets you drunk: ethyl alcohol (also called ethanol). Seems like an odd chemical coincidence, right? How could the same molecule responsible for something fun like getting tipsy *also* be responsible for ruthlessly killing germs on your hands? As with many chemicals, the answer is: it's all about the dose.

"Getting tipsy" is what happens when ethanol interferes with the normal functioning of our brains. Increase the dose, and you quickly transition from fun to loss of your gag reflex. (That's why you turned your college roommates on their side—so they wouldn't choke on their own vomit.) Increase the dose further and you're in overdose territory: coma and death are not out of the realm of possibilities. So it's actually not weird that the same molecule can have different effects at different doses. The weird part is that we actually *enjoy* the effects of widespread interference with our central nervous system.

Anywho, back to the point: How does ethanol kill germs?

And *does* it even kill them?

Yes, it does. But unfortunately, we don't really know how.

Why not? Well, it's fairly simple to design an experiment that goes something like this:

1. **Contaminate something with a known quantity of a specific type of germ (for example, one million cells of *E. coli*)**

2. **Apply a solution of, for example, 70 percent ethanol (and 30 percent water) for 30 seconds**

3. **Count how many live germs remain**

This experiment tells you how good a particular disinfectant is at killing a particular type of germ under a specific set of conditions. From a hundred years of experiments like these, we know that solutions of 60–95 percent ethanol are pretty damn good at killing a wide range of bacteria, viruses, and fungi. We also know ethanol is very *bad* at killing bacterial spores.* But overall, it's good at killing most of the germs that are likely to make you sick, including viruses. It's good enough, in fact, that the World Health Organization now recommends that healthcare professionals use alcohol-based hand sanitizer instead of washing their hands with soap and water (unless their hands were visibly soiled).†

So, experiments showing that ethanol works are plentiful and convincing. Unfortunately, they tell us nothing about *why* it works.

....

* We met spores back in chapter 3. They're bacteria in suspended animation that are tough to kill and can re-activate if you're not careful.

† This recommendation does not extend to consumers. Why? Healthcare professionals may have to clean their hands upwards of 100 times per day—so they need something that's fast, uber-convenient, and ultra-effective. Hand sanitizer fits that bill. We regular folk don't wash our hands nearly that often, and we're not constantly being exposed to infectious patients, so soap and water is just fine.

That's a different and much harder series of experiments. Why? Well, suppose you stumble upon a particularly grizzly murder scene. The victim's head is bashed in, all their limbs are cut off, their guts are spilled all over the place, to the point where you can barely tell that they were a person. It would be hard to figure out *exactly* what killed them. Was it blunt force trauma to the head? Was it blood loss? Was it a combination of the two? When ethanol kills a germ, it does so like a vicious crazy murderer: it goes for everything. It denatures proteins* (rendering them useless), and can dissolve cell membranes, causing germ 'guts' to spill everywhere. Which one of these injuries was the deadly blow? Tough to say, detective, tough to say.

So we don't know *exactly* how ethanol kills germs. But we know it does—at least under certain conditions. Remembering all the different conditions under which ethanol, or any other disinfectant, kills germs, can be overwhelming. To organize all of these variables, I find it helpful to write the process as a chemical reaction:

$$\text{GERMS} + \text{ETHANOL} \longrightarrow \text{DEAD GERMS}$$

Written like this, the questions almost ask themselves:

WHAT KIND OF GERM?   HOW MUCH ETHANOL?   HOW LONG DOES THIS TAKE?

$$\text{GERMS} + \text{ETHANOL} \longrightarrow \text{DEAD GERMS}$$

HOW MANY GERMS?

....

* Denaturation is the large-scale disruption of a protein's three-dimensional structure, which almost always renders it unable to do its job.

And the answers yield helpful guidelines to keep in mind whenever you're dealing with dirty hands.

First: What kind of germ are you trying to kill? Is it a virus? A bacterium? A fungus? As we've seen, ethanol is a pretty effective killer, except when it comes to bacterial spores. If you want to kill those, you'll need to use something else.*

Next: How many germs? If someone with the flu hacks up a lung directly into your palm, that's a lot more viruses in your hand than if you touched a contaminated subway handle for a split second. The more germs need to be killed, the more you have to consider the next two questions: How much, and how long?

How much? This is a tricky one. It can refer to how *concentrated* the ethanol solution is (which is measured as percent by volume, and is usually listed on the label). Ethanol is effective in the 60–95 percent range, so most hand sanitizers are 60–70 percent ethanol. Whiskey, vodka, and other hard liquors are usually less than 40 percent ethanol, so they won't kill enough germs to prevent you from getting sick. "How much?" can also refer to how much sanitizer you use. And that's somewhat tied to the next question: How long?

Whatever ethanol is doing to kill germs, it takes time. The longer you keep the ethanol in contact with the germs, the more germs it can kill. Unfortunately, alcohol evaporates pretty quickly, which you'll know if you've ever used rubbing alcohol on your hands. That's why hand sanitizers are usually formulated as gels, not free-flowing liquids: the gel prevents the alcohol from evaporating so quickly, allowing it more time to kill germs. (Plus, it's much easier to use a gel without making a mess.) Ethanol typically needs at least thirty seconds to do its job. So when you're using hand sanitizer, make sure you: (a) use enough to fully coat

....
* Unless you are in contact with someone who has, say, a *Clostridium difficile* infection, you probably are not exposed to dangerous bacterial spores.

your hands, and (b) don't wipe it off. Just keep rubbing, getting it in all the nooks and crannies, until all the alcohol evaporates.

You may also be curious about surface disinfectants—products you'd use on doorknobs, taps, countertops, etc. These are usually formulated without alcohol, because it's especially flammable when spread out over a large surface or sprayed as a fine mist. (No one wants to swap infectious diseases for house fires.) There are loads of different active ingredients used, and they can take wildly different amounts of time to kill germs—usually much longer than thirty seconds. That's why it's *critical* to actually read the directions on the back of any product that claims to kill viruses or bacteria and then *actually follow those directions*. Usually, the directions on a spray or wipe product tell you to fully wet the area with disinfectant (to make sure there's enough to kill germs) and then let it sit undisturbed for 4–10 minutes (to make sure the disinfectant has enough time to do its job). If you skimp on either *how much* or *how long*, you're probably not getting the full protection claimed on the label.

There's one more implicit question not addressed in our equation above, and it applies to both hand and surface disinfection. If your hands or a surface are "visibly soiled" with dirt, oil, or any bodily fluids, including but not limited to pee, poop, blood, snot, vomit, spit, etc., you should *not* use hand sanitizer or a disinfectant. Instead, you should thoroughly wash with soap and water.

Why?

Let's look specifically at hand sanitizers. We don't know *exactly* how ethanol kills germs, but we do know that it denatures proteins: it transforms them from their native, functional form into shapeless, useless globs that don't do their jobs and stick to each other. Most scientists believe that protein denaturation and coagulation is at least partially responsible for ethanol's germ-killing ability. So let's take our earlier chemical reaction and flesh it out a bit:

$$\text{PROTEINS IN GERMS} + \text{ETHANOL} \longrightarrow \text{DENATURED/COAGULATED PROTEINS IN GERMS}$$

Now let's suppose that you've got crusted blood all over your hands. That blood contains a *ton* of protein. Our chemical reaction now looks like this:

$$\text{PROTEINS IN GERMS} + \text{PROTEINS IN BLOOD} + \text{ETHANOL} \longrightarrow \text{?}$$

Instead of just the protein in the germs, there's also all the protein from the blood. Ethanol doesn't know to ignore the blood proteins and just focus on the germ proteins. It'll react with both, and that means there's less ethanol available to denature germ proteins. Bottom line: proteins in blood (and potentially other molecules in other types of dirt, grime, or bodily fluids) can shield proteins in germs from the sanitizer. So, if your hands have any gunk on them, wash thoroughly with soap and water. Ditto for a surface: if there's congealed gunk, it can actually protect germs from the disinfectant. Soap and water (and the scrubbing that goes along with it) can break up all that gunk and wash away the germs.

Lastly, because "gets-you-drunk-ethanol" and "sanitizes-your-hands-ethanol" are one and the same molecule, you might be wondering whether you could get drunk off hand sanitizers, either by drinking them or by rubbing so much onto your hands that the alcohol could seep through your skin and end up in your bloodstream. On the first point, hand sanitizer manufacturers typically add bad-tasting chemicals called "bitterants" to strongly discourage you from drinking it. (Some sanitizers are based on isopropanol, a different alcohol that does not get you drunk.) On the second, there have actually been a couple of studies done. In the largest, twenty healthcare workers took ultra-sensitive breathalyzer tests and had their blood drawn, then used hand sanitizer thirty times in a one-hour period, then redid the tests. Ethanol *was* actually detectable in some of the twenty, though at super low levels—levels that would be undetectable by a standard roadside breathalyzer test. So there ya go.

ACKNOWLEDGMENTS

To my editor at Dutton, Stephen Morrow, thank you for turning this book from something only a mother could read into something my mother will actually read.

Mom and Dad, this book would not exist without your love, care, support, generosity, and eternal optimism. Thank you.

Julia, you're the shit. I fucking love you. Also, thanks for not breaking up with me. Miguel, dyno to the crimp! Thanks for pushing me up the wall of life. Pascale, Carine, and Muriel, remember dressing up in all black and sneaking into our own houses? I still do that. Wojtek and Ricky, one day I swear I'll golf again. Thanks for all the free therapy. Kenny, the draft you're reading is a little out-of-date. Kemps, it's an honor to be an honorary K. Kels, and Christina, next time I have ketchup on my face, please tell me. Andrew, thanks for your friendconomist review. Claudia, anytime you need electronics shipped overseas, I gotchu. Dan, can't wait to read *your* book! Nina, thanks for the encouragement and mountain pictures. Waseem, Siggi, and Laika, I'm looking forward to Friendship Lane. Tony and

Patricia, many of these pages were written under your roof and with your encouragement and support. Looking forward to more gardening! Nocci, thanks for all the licks.

Elizabeth Choe: thanks for helping me figure out what *Ingredients* was, before it was anything. Thanks to James Williams, Danielle Steinberg, and the rest of the team at *National Geographic* for helping Elizabeth and me produce the series. (If you want to do a second season . . . now's probably a good time!) Susan Hitchcock started this whole "write a book" thing by inviting me to pitch at *NatGeo*, and then introducing me to her—and now my—agent, Jane Dystel.

Jane's e-mails provide a behind-the-scenes peek into the genesis of this book:

JUST CURIOUS

CHECKING IN

CHECKING IN AGAIN

I DO NEED TO TALK WITH YOU

I REALLY NEED TO TALK WITH YOU

OFFER!!!!

Jane, seriously: thank you.

This book would not have been possible without the incredible generosity of Sue Morrissey, Glenn Ruskin, Dave Smorodin, Flint Lewis, and Human Resources at ACS, who let me take six months off to write this thing . . . and then let me come back. To Hilary Hudson, who bore the burden of ACS's generosity, thanks for running the ship while I was gone. And to the rest of the *Reactions* team: thanks for correcting my mistaken impression that Stevie Nicks is a basketball player for the New York Knicks. All views, opinions, and bad puns in this book are my own and not those of ACS.

Caitlin Murray, you prevented me from stepping on several

rakes, and your perceptive and thoughtful criticism would have been especially useful if I had sent you the draft months before I did. Thanks for never barring any holds. Douglas, I'll show *you* an asshole. Christine Ball and John Parsley, you assembled a crack team that made Dutton the perfect home for this book. Hannah Feeney, thanks for the notes, and answering my 872 stupid questions with the patience of Job. Kaitlin Kall, we've never met, but the moment I saw the gigantic Cheetophallus on the cover, I felt the sense of humor of a fellow traveler. Lorie Pagnozzi, you turned eleven-point Times New Roman into a stunningly readable interior. And David Chesanow, I know we disagree about commas, but you caught all my embarrassing Disney-related mistakes, so I suppose we can postpone our duel. (And I'll never misattribute "A Whole New World" again.) Joel Breuklander, Susan Schwartz, and Leila Siddiqui: thank you. To the Penguin Random House Legal Team: thanks for the indemnification. And sorry about the Cheeto penis. Fellow Dutton author and former neighbor Daniel Stone, thanks for telling me what to expect, and for the whiskey. John Essigmann, you had much better things to do than comfort a college kid going through a breakup at 1:00 A.M., but you woke up and did it anyway. MIT is a warmer, friendlier place with you in it.

Many, many scientists read excerpts of this book and pointed out errors or provided additional context. Regina Nuzzo was a statistical sensei extraordinaire; her time and energy resulted in an immeasurably better Part III. Jay Kaufman was the epidemiological conscience. Alyson van Raalte and Michal Engelman were demographic deities. John DiGiovanna was a solar superhero. Dennis Bier, call anytime to vent. Tyler VanderWeele, thanks for letting me sit in on your class. Katherine Flegal, thanks for picking bones. Walter Willett was impeccably cordial and warm even though he probably disagrees with 90 percent of what's in this book. Dylan Small, thanks for letting me crash your causal inference party. David Jones, thanks for assigning all those essays back in 2006. Cherie Pucheu-Haston, thank you for those incredibly detailed e-mails. David Spiegelhalter, I'm really sorry I kept you waiting for twenty minutes after our scheduled interview time. I honestly have no clue how that happened.

Many others were extremely generous with their time in the name of factual accuracy: Ken Albala, David Allison, Philippe Autier, Charlie Baer, Ray Barbehenn, Bob Bettinger, Doug Brash, Dan Brown, Kelly Brownell, Vincent Cannataro, David Chan, Peter Constabel, Alyssa Crittenden, Jennifer DeBruyn, Patti deGroot, Brian Diffey, Joanna Ellsberry, Scott Evans, Cree Gaskin, Chris Gardner, Ros Gleadow, Sander Greenland, Gordon Guyatt, Kevin Hall, Bill Harris, Stephen Hecht, Melonie Heron, Missy Holbrook, Casey Hynes, John Ioannidis, Gulnaz Javan, Nishad Jayasundara, Lene Jespersen, Tim Johns, Chantal Julia, Martijn Katan, David Klurfeld, Susanne Knøchel, Kristine Konopka, David Kupstas, Tracy Lawson, Bill Leonard, James Letts, Yanping Li, Lucy Long, David Madigan, Ramsey Markus, Fabian Michelangeli, Carlos Monteiro, Leif Nelson, Laura Niedernhofer, Brian Nosek, Sam Nugen, Betsy Ogburn, Uli Osterwalder, Chirag Patel, Tom Perfetti, Austin Roach, Andreas Sashegyi, David Savitz, Leonid Sazanov, Rodney Schmidt, Katia Sindali, Kat Smith, George Davey Smith, Bernard Srour, Janez Stare, Vas Stavros, Dawnie Steadman, Michael Stepner, Diana Thomas, Bob Turgeon, Peter Ungar, Lee-Jen Wei, Bob Weinberg, Forest White, Torsten Will, Adam Willard, Sera Young, and Stan Young.

And, last but not least, to the people who helped me become who I am, thank you: Bassem Abdallah and Hilary Bowker, Maggie Abu-Fadil Chiniara, Samira Daswani, Alex Frank, Max Hunt, Tara Nicholas, Mike Rugnetta, Gabriel Sekaly, Alex Snider, Lisa Song, and Amandine Weinrob.

SOURCES

Rather than waste a bunch of paper printing out every single source I used, I'm wasting server space instead. Head on over to www.ingredientsthebook.com for the full list of sources, including links to the original papers. And remember, if you think I screwed something up, shoot me an e-mail at oops@ingredientsthebook.com and I'll dig into it.

INDEX

ABOUT THE AUTHOR

George Zaidan is a science communicator, television and web host, and producer. He created *National Geographic*'s web series *Ingredients: The Stuff Inside Your Stuff*, and he cowrote and directed MIT's web series *Science Out Loud*. His work has been featured in the *New York Times*, *Forbes*, the *Boston Globe*, *National Geographic* magazine, NPR's *The Salt*, NBC's *Cosmic Log*, *Science*, Business Insider, and Gizmodo. He is currently the executive producer at the American Chemical Society. *Ingredients* is his first book.